Ethnicity, Identity and Faith in the Current Migratory Crisis

Roberta Ricucci

Ethnicity, Identity and Faith in the Current Migratory Crisis

Continuity and Change in Migrants' Religiousness in Southern Europe

Springer

Roberta Ricucci
Department of Culture, Politics and Society
University of Turin
Turin, Italy

ISBN 978-3-030-84055-6 ISBN 978-3-030-84056-3 (eBook)
https://doi.org/10.1007/978-3-030-84056-3

This Springer imprint is published by the registered company Springer Nature Switzerland AG
The registered company address is: Gewerbestrasse 11, 6330 Cham, Switzerland

Preface

Year after year, everyday life in Europe is increasingly intertwined with the lives of citizens from other countries.

A powerful factor in social transformation, the "useful invading" migrants have introduced customs and traditions to the places where they settle, often through the history of several generations, into the social structure of the country of arrival until they reach the top step, that is to become citizens and be able to be mistaken for natives. In 2016, immigration was in fact considered the main problem by about 50% of Europeans (compared to 14% in 2010) (Papademetriou & Banulescu-Bogdan, 2017). Moreover, in subsequent years, the situation did not change, according to data from polls and media coverage (De Coninck et al., 2019). However, many elements told of a Europe that achieved a peaceful and constructive coexistence, of inter-culture, of the fruitful encounter among points of view, cultural traditions and ways of doing things. As explained by those who studied the mechanisms of news of events, these issues did not make headlines. Their story could be of benefit to citizens in general, from small and large realities, in order to dampen positions that were often spoiled by clichés and stereotypes (Crawley & Skleparis, 2018; Dennison & Geddes, 2019). In other words, to break down at least those walls built on stereotypes and prejudices.

Although the European Union has been able to face—albeit with great effort and social costs—the global economic crisis, employment and the deep social implications of recent decades and nationalistic movements, being an EU citizen, or not, can translate into advantages or limitations in international mobility, in beneficial opportunities, in protection that can be accessed. Such general considerations find their concretization in the daily life of each person, in the warp of relationships, contacts, meetings (and clashes): it is here that, in addition to the problems of a bureaucratic and regulatory nature, a passport may not be enough.

In the history of the migrations of Europeans to America or Australia in the twentieth century, integration meant economic, cultural and social assimilation to be crowned with entry into the community of citizens. For some time now, the biographies of migrants have been written by interweaving old and new affiliations, in which transnational dynamics are strengthened through the generations. The hypothesis of a transnational or post-national citizenship is gaining ground, due to the size of the migratory flows and the reluctance of many immigrants to naturalize. But also the unwillingness of some European countries to accept that immigrants have become permanent residents and the refusal of as many countries of emigration to abandon the rights of their citizens abroad. Today, more than yesterday, under the label of migrant or foreigner, non-EU citizen or resident, there are biographies, stories of departures and arrivals, of different and complex legal conditions, characterized by multiple (invisible) barriers to the full exercise of social, civil and political rights. The expansion and changes of migration over the last few years—in particular the aspect of the acquisition of citizenship—have brought the issue of political integration under the spotlight at global level.

In 2000, the European Commission launched the programme "Building an Inclusive Europe" and the European Social Fund was the instrument that, most of all in terms of investment, sought to help the various Member States to invest in training, employment, equal opportunities and the fight against discrimination.

In other words, to build inclusion policies in a European area that faced, among others, the challenge of becoming an increasingly attractive destination for migratory flows, which have been growing stronger from generation to generation.

At the beginning of the new millennium, Europe was responding to increasing globalization with the challenge of the knowledge society and social inclusion, a long and complex process that called for policymaking, as well as the allocation of specific funds. On a daily level, it translated into the fight against the processes of exclusion and marginalization, in the promotion of diversity. If, strictly speaking, social inclusion could be defined as the full realization of citizenship rights, we could understand it in the broader dimension of overcoming situations of social closure.

The need to understand what is conducive to social inclusion and what is being done in this direction has highlighted how we are struggling on the European continent. The litmus test of this difficulty is emblematically the confrontation with migrant citizens and their diversity. Foreigners are a stable presence, which takes root and looks to the future in the local, national and sometimes European horizon. Faced with this scenario, confirmed by the data on entry, births and migration seniority, the management of the "foreign presence" seems still reduced only to border control issues (who to let in and under what conditions) and profiles (economic migrants, asylum seekers and refugees, highly qualified).

Recalling King (2015), one could say that, "it can also be argued that, in contrast to the past – either the nineteenth century or the early post-war decades – we live in an age of migration control. So that, within the scenario of a 'world on the move', there are also powerful forces preventing international migration, and causing a state of forced immobility in the age of migration" (….). Indeed, "migration is an eminently political topic. Over the past decade, the politicization of migration has been evidenced by a series of developments: the fear in Western countries of an influx of masses of migrants from countries of the former Soviet bloc, and in European Union countries of an invasion by citizens from new member countries with each enlargement of the Union; the questioning of the role of migrants in the economic and social upheavals triggered by the financial crisis in South-East Asia; restrictive policies and anti-immigration backlash in the wake of the terrorist attacks of 11 September 2001; renewed outbreaks of xenophobia in several African countries that blame domestic crises on migrants; and the exploitation of migration issues by some politicians to gain electoral mileage. All these examples illustrate the close links between economic, political and social issues on the one hand, and mobility on the other. More than ever therefore, migration is a ready target with psychological, economic and public relations connotations" (OIM, 2003).

With these words, in 2003, the IOM (International Organisation for Migration) opened its report on migration. Fifteen years later, the situation did not seem very different. Certainly, some protagonists changed, the crisis in South-East Asia was accompanied by the crisis started by the B & L in 2008 and the issue of Islamic terrorism was no longer (unfortunately) limited to 11 September. Today, as yesterday, however, migrants are the subject of political debate, due to their characteristics: in other words, they are concerned about the cultural diversity and the mix of biographical paths, languages, religions, traditions and values that many authors have instead shown to be the engine of development.

In Italy[1] and throughout the European Union, the public debate on immigration has progressively intensified since the years of the recent economic crisis, with its heavy consequences on the employment and income prospects of the population, and taking into account the rapid increase in the number of asylum seekers in recent years.

Also today, as yesterday, therefore, the relationship with otherness is an unresolved question; debates and data, political positions and legal measures that attempt to exclude, limit and contain it are national examples. Italy has been fully involved in this debate for some time. In the following sections, we will try to highlight some features of the national debate on migration.

Religious Identity Matters

Las Ramblas in Barcelona. Two young women, one with a headscarf, the other without. I observe that they are approached by a group of youths who make fun of them because of the headscarf. More girls come to their defence. The many tourists show no interest.

Rome. Extreme periphery. A boy comes out of a courtyard after praying in a place of worship in an old workshop, a kind of "improvised mosque". Two peers on a motorcycle insult him, "Go back to your own country! There is no place here for you Muslims".

[1] The text takes up and expands the contents of Ricucci (2018a, b, c).

Berlin, crossing the Tiergarten park. A group of young Sikhs has improvised a game of cricket. I look around. Nobody is paying attention. Not far away, a group of children is playing football.

Vignettes of cities transformed by successive immigration waves, where religious pluralism has become part of shopping streets (Barcelona), the periphery (Rome) and Sunday normality (Berlin).

In the post-2001 climate, reinforced by several episodes of violence inside and outside Europe, religion has returned to the agenda of European governments, even those most attached to the principle of secularism.

Religion cannot be separated from the rest of life contexts, the socio-cultural system in which we grow up, which nourishes us daily through old and new media. Every day the diet we partake of from newspapers, television and the web is rich in religious imagery from Game of Thrones to Lucifer to Unorthodox, from Instagram tattoos to films such as *Habemus Papam* or to the series on God with Morgan Friedman (Bradatan & Ungureanu, 2014). In images of mosques or in those of Boko Haram, the use of sacred symbols and the fight in the name of some God are part of everyday reality in a society which many had considered secularized and which—even in France, the home of secularity—continues to be studded with references to religion under the spotlight—not only because of terrorist attacks but also because of full churches, the spread of spiritual groups, the seeking out of sense, which leads many young people to approach the sacred, all giving the lie to a prophecy that seemed indissolubly linked with advanced modernity.

While religion is present in the lives of people and, surprisingly, of many young people, its forms and how it is lived are adapted to, and transformed by, modernity and by the local and/or transnational community, family and individual experiences implied by a migratory path (Phalet et al., 2011; Van der Bracht et al., 2013).

It is easy for misunderstandings to creep in or space opened for stereotypes, which—in a period of fear of otherness—may be harbingers of conflict.

Yet attempts at reinforcing often only weaken the armour constructed to understand (or survive in) an increasingly complex society. One ends up adopting a synecdoche, a part of the whole, which is to say using religion to explain the complex identities, which each one of us possesses. This leads to a so-called "sacralisation of society", a paradox in Europe where for a long time the secularization thesis has held sway. Religion becomes a potent synecdoche to explain events without taking anything else into account. Therefore, it happens that "in the newspapers, in the media, deaths are attributed to religious extremism if not simply religion. In rare articles and a few books, economic, political and social explanations are put forward". Is it important to articulate and integrate different motivations in order to understand and attack? Is there "any difference whether the explanation is social or religious? As distinct from the religious, social explanations allow us to distinguish, to check, to compare situations as well as the convictions of the groups and individuals involved. They do not oblige us to contrast 'us' long-term Europeans who cannot help calling ourselves Christians, which we are not in reality – we go to bars and concerts, practise sexual freedom and gender equality, with 'them' who are

Muslims i.e. subjected, force their women to wear the veil, practise polygamy and kill anybody who does not share their faith" (Ciafaloni, 2016, p. 32).

In a 2015 article, the French political expert Olivier Roy helped to comprehend what for many was incomprehensible. To paraphrase his words, "We have welcomed the Muslims, given them our jobs, assisted them with social welfare and opened our schools to their children – and this is how they pay us back, by blowing us up, killing our women and children".

Recent opinion polls in Europe reveal that the above sentiment is widespread on the continent. The French sociologist explained clearly that it was not the sons of Islam as such who were tainted by extremist behaviour. It was those who had never been taught a "culturally integrated" Islam, or who felt isolated, trapped between a consumer society who had never fully accepted them, rather discriminating against them on the grounds of their origins, and their community of origin which condemned them because of their "non-Halal" behaviour.

The Need of Being Focalized at Local Level

Assorted research (Zincone et al., 2011; Entzinger & Scholten, 2015) discredited the idea of the effectiveness of national models of integration in the European context and their unchangeableness at local level, so that attention shifted towards territorial scenarios where scholars attempted to understand in depth the dynamics that sprung up among natives, migrants and second generations. Literature on the great models into which various ways of incorporating immigrants were subdivided (from assimilationism to multiculturalism, from *Gastarbeiter* to differentialism) put us on our guard, however, against the illusion that national policies no longer played an important role in intervening on the insertion paths of non-autochthonous citizens. At the same time, it warned us not to ignore (as sometimes happened in the past) the influence of the local context and everyday dynamics "triggered by diversity in the management of daily life" (Vertovec & Wassendorf, 2005, 2010). In this framework, management of immigration in Italy offers various examples of how national norms become declined differently at local level.

The game of inclusion and social cohesion (of which religion is an important theme) is really played out on the field of integration policies at local level, including relations with Muslims, both old and young, first and second generations. Obviously, towns, cities and neighbourhoods differ regarding the composition of the immigrant population, its socio-demographic characteristics, initiatives and opportunities available to them, and the modalities of relations with the general citizenry and consequent social-cohesion processes. In other words, they differ as to so-called "local integration policies" and local management of religious diversity. Among these strategies, there are those pertaining to the management of religious difference, which—albeit in a multireligious environment—remind us of relations between towns/cities, and the Muslim presence. Among immigrants, religiosity is alive, as evidenced by the existence of mosques, churches and prayer halls. Of course, these are not only places of worship but are also welcoming centres offering a kind of social support system both for new arrivals and the irregular immigrants (Portes & Hao, 2002). On the other hand, migrants feel at ease in places of worship—there, they find clergymen speaking their native language, sharing (or at least understanding) their cultural and ethnic background, aware of the difficulties of the meeting/clash between familiar ways of life and those expected by the host society.

To some extent, the local dimension has been overlooked in that it means, on the one hand, that local policies intervene in managing the daily lives of Muslim communities and, on the other, that it is the arrival on the scene of the second generation that modified (strengthening or weakening, modifying or erasing) the demands for recognition by their fathers and mothers, which sometimes provoked a public reaction on the part of the citizenry.

Going Beyond the Status Quo: Changes in Ethnicities, Generations and Ways of Believing

A diachronic analysis of the data on migration flows in Europe underlines how the origin of migrants has moved progressively first towards Latin America and more recently Eastern Europe, with migrations from Poland, Romania, Ukraine and Bulgaria becoming more and more relevant since the late 1990s (Zahra, 2016).

Yet, despite the affirmation of increasing religious pluralism, research in Europe concentrated on the growing Muslim presence (Hunter, 2002; Cesari, 2014), with observations on and studies of different viewpoints, as mentioned above.

At the same time, immigration countries in Europe are now facing a change in their immigrant population: thanks to the migration waves from Central and Eastern Europe, an increasing number of Catholics (and other Christians) are now visible on the migration scene (Casanova, 2009). On the other hand, flows arriving as asylum seekers are re-framing the ethno-religious composition.

Consequently, considering the evolution of the migratory process, the almost exclusive focus on the Islamic component allowed little investigation of the increase in the Christian component, following migrations from Eastern Europe or from South America, as well as studies of other religious organizations (Modood & Triandafyllidou, 2012).

Furthermore, the youth perspective and the religious point of view were also under-represented in studies on the European side of the Atlantic[2] (Foner & Alba, 2008). For these, the processes of social integration, as well as the definition of cultural identity, in many instances, were filtered through the relation with religious institutions and the educational socializing processes, and through the support they

[2] In US sociological research, increasing attention has been paid to the relations between Latino second generations and both their religious belonging (and conversion to Evangelicals and Protestants) and religiousness (Rumbaut, 1997; Crane, 2003).

offered (Ricucci, 2014). Families, especially mothers, gave religious education to children: a minority was defined as "convinced and active believers" but the great majority of young people were generally defined as "believers from tradition and training", where religion played a strong social role (opportunity to meet with peers and use the mother tongue). In a way, they were religious for social reasons instead of religious purposes. This applied to both Christians and Muslims.

Speaking about children of immigrants, relations with religion needed to be framed in two different perspectives. On the one hand, there was the acculturation process, i.e. how they fitted into the receiving society (Portes, 1996; Berry, 2011); on the other hand, how they defined their identity as young people with a migratory background in a globalized world (Crul et al., 2012; Voas & Fleishmann, 2012).

The ways in which the cultural identities of immigrants change over time and under the influence of the larger society, as do the characteristics of the relationship between them, are generally crucial issues in the study of the integration process of second generations. For children of immigrants rejoining their parents who emigrated before them, this process is more complex because it overlaps the larger development task of identity formation, which started in their home countries and then continued in the receiving society. The outcome of this process mirrors the various possible acculturation attitudes. Adolescents and young people can choose among various options: if they do not solve the cultural identity issues that they face, they may exhibit identity diffusion. They acquire within the family the adaptive patterns of behaviour, personal characteristics, values and social responses expected of them in their cultural heritage. This is why even the issues of religiousness, religious belonging and the way of being religious need to fit at the crossroads of various sociological (and psychological) disciplines. Indeed, as Berry (2006) demonstrated, there is clearly evidence of the importance of religion in how immigrants engage in both their intercultural relations and their integration paths. Moreover, specifically, Christians earn the highest integration score.

References

Berry, J. W. (2006). Acculturative stress. In W. J. Lonner & R. S. Malpass (Eds.), *Psychology and culture* (pp. 211–215). Allyn and Bacon.

Berry, J. W. (2011). Immigrant acculturation: Psychological and social adaptations. In A. Assaad (Ed.), *Identity and political participation* (pp. 279–295). Wiley.

Bradatan, C., & Ungureanu, C. (Eds.). (2014). *Religion in contemporary European cinema. The postsecular constellation*. Tayor & Francis.

Casanova, J. (2009). Immigration and the new religious pluralism: A European Union – United States comparison. In G. Brahm Levey & T. Modood (Eds.), *Secularism, religion and multicultural citizenship* (pp. 139–163). Cambridge University Press.

Cesari, J. (2014). *The handbook of European Islam*. Oxford University Press.

Ciafaloni, F. (2016). *Integrazione e differenziazione*. Comitato oltre il razzismo.

Connor, P. (2009). International migration and religious participation: The mediating impact of individual and contextual effects. *Sociology Forum, 24*, 779–803.
Crane, K. R. (2003). *Latino curches: Faith, family and ethnicity in the second generation*. LFB Scholarly Publishing.
Crawley, H., & Skleparis, D. (2018). Refugees, migrants, neither, both: Categorical fetishism and the politics of bounding in Europe's "migration crisis". *Journal of Ethnic and Migration Studies, 44*(1), 48–64.
Crul, M., Schneider, J., & Lelie, F. (2012). *Second generations in Europe*. Amsterdam University Press.
De Coninck, D., Matthijs, K., Debrael, M., De Cock. R. & d'Haenens, L. (2019). Unpacking Attitudes on Immigrants and Refugees: A Focus on Household Composition and News Media Consumption. *Media and Communication, 7*(1), 43–55.
Dennison, J., & Geddes, A. (2019). A rising tide? The salience of immigration and the rise of anti-immigration political parties in Western Europe. *The Political Quarterly, 90*, 107–116.
Foner, N., & Alba, R. (2008). Immigrant religion in the US and Western Europe: Bridge or barrier to inclusion? *International Migration Review, 42*, 360–392.
Hunter, S. T. (Ed.). (2002). *Islam, Europe's second religion: The new social, cultural, and political landscape*. Praeger.
King, R. (2015). Migration comes of age. *Ethnic and Racial Studies, 38*(13), 2366–2372.
Modood, T., & Triandafyllidou, A. (Eds.). (2012). *European multiculturalisms. Cultural, religious and ethnic challenges*. Edinburgh University Press.
OIM. (2003). *World migration report 2003*. OIM.
Papademetriou, D., & Banulescu-Bogdan, N. (2017). *Understanding and addressing public anxiety about immigration*. MPI.
Phalet, K., Güngör, D., & Fleischmann, F. (2011). Religious identification, beliefs, and practices among Turkish-Belgian and Moroccan-Belgian Muslims: Intergenerational continuity and acculturative change. *Journal of Cross-Cultural Psychology, 42*(8), 1356–1374.
Portes, A. (1996). *The new second generation*. Russell Sage Foundation.
Portes, A., & Hao, L. (2002). The price of uniformity: Language, family and personality adjustment in the immigrant second generation. *Ethnic and Racial Studies, 25*(6), 889–912.
Ricucci, R. (2014). *Second generations on the move in Italy*. Lexington Books.
Ricucci, R. (2018a). Arrivare non è tutto. Barriere invisibili nel passaggio da migranti a cittadini. In G. Borgognone, D. Carpanetto, & L. Segreto (Eds.), *Dalla caduta del Muro ai nuovi muri. Trasformazioni globali e scenari internazionali 1989-2018* (pp. 311–321). UTET.
Ricucci, R. (2018b). *Cittadini senza cittadinanza. La questione dello jus soli*. Seb27.
Ricucci, R. (2018c). Giovani stranieri cristiani in Italia: Una minoranza invisibile che interroga i luoghi di culto etnici. *Religioni e Società, 90*, 46–56.
Rumbaut, R. G. (1997). Paradoxes (and orthodoxies) of assimilation. *Sociological Perspectives, 40*(3), 483–511.

Van der Bracht, K., Van de Putte, B., & Verhaeghe, P. V. (2013). God bless our children? The role of generation, discrimination and religious context for migrants in Europe. *International Migration, 51*(3), 23–37.

Vertovec, S., & Wassendorf, S. (2005). *Migration and cultural, religious and linguistic diversity in Europe: An overview of issues and trends*. Compass.

Vertovec, S., & Wassendorf, S. (Eds.). (2010). *The multiculturalism backlash. European discourses, policies and practices*. Routledge.

Voas, D., & Fleishmann, F. (2012). Islam moves west: Religious change in the first and second generations. *Annual Review of Sociology, 38*(1), 525–545.

Zahra, T. (2016). *The great departure: Mass migration from Eastern Europe and the making of the free world*. W.W. Norton & Comp.

Zincone, G., Penninx, R., & Borkert, M. (2011). *Migration policymaking in Europe: The dynamics of actors and contexts in past and present*. Amsterdam University Press.

Turin, Italy Roberta Ricucci

Acknowledgements

Working with migrants and studying the various facets of mobilities is part of my career. I remember my first summer camping with a group of unaccompanied minors in 1999 in Liguria (Italy). It was an adventure, full of mixed feelings: crossing roads and being at the sea with a group of Moroccans and Albanians to look after together with a Cameroonian, both of us volunteers meant being under observation by people. At that time, it was quite unusual in Italy to see foreign teens play and have fun instead of being "vu cumprà".[1] Ladies clutched their bags; men commented aloud that they would call the police to clear out a dangerous group of foreigners. In fact, 15 boys between 15 and 17, with 5 animators, a priest and a nun. A group of subversives, no doubt. It was useless to explain the goal of that camping trip: offering a week's holiday to minors without parents. Their skin colour and language were the two (and only) variables people were aware of. Later, maybe, the same ladies had a Romanian caregiver for their old parents and the men a Moroccan or Senegalese blue-collar worker in their small firms. Perceptions and stereotypes matter a great deal. This kind of attitude is transversal across gender, social class, level of educational capital, generations, nationalities and religious belonging.

Over the last two decades I have been in touch with many passionate and wonderful people who have stimulated me to think about the various facets of migrations and their protagonists. Among them there are numerous volunteers, associations and institutions with a religious background. Sometimes it is explicit: Caritas or Jesuits, Waldensians and an association such as Islamic Relief are places which are immediately linked to religion. In several cases, however, the link is veiled: it could concern the volunteers' values or the memories of a family socialization; it could be found in the name of an association, without any reference in the working attitude of its members. One thing is clear; the link between migration, religion, acceptance and inclusion is clearly present in my study and voluntary work with children and adults of foreign origin. First it was the unaccompanied minors, then the children of

[1]This was an expression referring to African immigrants during the 1980s who used to sell goods on the streets.

foreign families, and more recently asylum seekers. The activities carried out directly with them or with operators and volunteers are part of a wealth of precious experiences for those who want to study and deepen the real meaning of being a migrant today. For this reason, my first thanks go to them, coming from very different countries and with biographies and stories of resilience that often make you shudder. To these who wanted to share with me dreams and fears or hopes and difficulties.

Every discussion, every suggestion and every interaction with people working in the various social fabrics across Southern Europe and every debate at meetings, lectures, symposia and workshops with colleagues has impacted on and shaped this book. And so I must begin by acknowledging them, even if I cannot list every single person by name.

I would also like to thank all my respondents for generously letting me share their world, for their immense generosity and support: in particular several volunteers and social workers who involved me in a lot of their daily activities, giving me the opportunity to enter reception centres, associative activities and families without disturbing them, to participate in community activities and (sometimes) debates, to enjoy dinners and share religious holiday moments.

I also would like to thank the local institutions (and especially the Turin one), associations and organizations which provided me with useful contacts, to some extent validating me through their mediation.

A special thanks to my key informants for their human and intellectual support, not forgetting my translators for their proficiency, patience and kindness.

Introduction

July 2017, a dream fulfilled. I finally managed to visit the Montreal stadium where the Romanian gymnast Nadia Comaneci scored a perfect 10 in the 1976 Olympics. At the same moment, an event occurred which linked a child's dream with a sociologist present. In a wing of the sports palace, they were fixing up a welcoming centre for asylum seekers and refugees. A temple of sport, used to honour De Coubertin's values, was being adapted to provide a home for those who had been forced to abandon theirs. I tried to grasp the situation, discreetly watching what was going on: volunteers, young and old, religious and lay, were actively preparing what for many Syrians would be a new beginning.

Approaching the volunteers, I discovered that they were Canadians of Italian or Portuguese origin, or recent South American or Asian immigrants. There were Catholics, and other Christians and Muslims, who remembered that migrations were part of human history and therefore part of the DNA of every one of us, independently of languages, customs or religions.

Distinctive or common traits, which may have been used at every latitude to stigmatize or to facilitate insertion.

A year later in Barcelona, Spain, I found myself discussing interreligious dialogue with a group of youths who described themselves as "Europeans with an immigrant background". Carlos, Rachida and Suad were second generations involved in intercultural associations, which also paid attention to religion. They said they could not find room in parent associations or others already existing. Even from the cultural and religious point of view, they labelled themselves "a new generation", different from how the media portrayed them. They were active, assertive and pungent in their desire to be autonomous protagonists expressing their multiple identities.

I found a similar mood in Turin, Italy, where for years a group of young people, mainly Moroccan and Egyptian, had organized the "Open Mosque" exhibition. The intention was to make Islam and its spaces known to residents who—from fear or laziness—were guided by stereotypes. Last year they won an EU prize, from a network of Municipalities and Third Sector subjects, for "The Best European

Intercultural Initiative". They too were second generations, many of them university graduates, active citizens interested in promoting a European Islam.

These are examples of a research path aiming to focus on the relationship between religion and immigration from the various actors' perspectives: migrants (especially youth) and European citizens. In some way, all are involved in dealing with asylum seekers and refugees, whether as principal actors or as welcomers. Some feel guilty and others take their distance because "It was exhausting to gain acceptance, and I don't want to relapse and again be considered a foreigner, a migrant. Do you think I'm racist? Anybody who has not experienced the insults, the spitting, the feeling of always being under the microscope because you're black, because they think you're Muslim, because you speak a language they do not know… if you have not gone through all of that you cannot understand or judge how we – the people of the rubber dinghies, the minibuses and the cars – behave" (Abduhllahi, Syrian, political refugee).

The young people's point of view captures best the dynamics and the processes through which the religion-migration link can forge, modify and condition professional and existential life stories.

Research Orientation and Methods

Every research into migrants has to deal with the social and political representations of the phenomenon, above all during the years of writing this book. The theme of migrations has assumed a central position in political debate whenever various aspects of society (housing, employment, education, welfare) interweave with the social costs of migrations, tending to polarize reactions. Dangerous classifications and identifications, enflamed by the populist right wing, are the corollary to the—scientific or non-scientific—debate on the subject. Stereotypes and commonplaces can be found in the comments of social operators and administrators, reinforcing a labelling process which has deep roots. Bearing this in mind, I spent days in intercultural and multicultural meeting places, regularly recording or transcribing conversations with participants. Visiting families, attending parties and rites of passage, was also most important. I attempted interviews, but it was mostly a matter of informal conversations with parents and relatives of foreign-origin adolescents and youths as well as children's entertainers, educators and other subjects who might be able to enlighten me about these young people's family background and future prospects. Thus, for my research I used an ethnographic approach and participant-observation techniques. My methodology required entering private (family) and religious (from prayer halls to temples, churches to mosques) environments.

The volume is mainly concerned with Italy, which, for historical and social reasons, is considered an exemplary country of the broader European context. From being a country of emigration, over the past three decades it has become the destination of significant migratory flows from both outside and inside the European Union. In a short space of time, the population of immigrant origin has become an important structural part of Italian society (about 10%). At the same time, this population has marked characteristics of quite homogeneous territorial diffusion and heterogeneity of nations and cultures of origin.

For all these reasons, Italy is therefore the main country considered: it could be a kind of "case study". The idea of a permanent or temporary abandonment of the country has returned. Some people say that it has been an ever-present element of Italian identity.

Taking into account Italy's recent history, the scale of youth emigration is significant. By proposing a methodological and theoretical perspective of biographically oriented sociology of youth and the sociology of mobility, I would like to formulate some questions and point out some problems. Among others, to what extent does the fate of youth intra-EU have a typical character, common to this social group, and to what extent is it exceptional and unique compared with other kinds of migration? Youth mythologies make us believe in the uniqueness of their experience. Is it then possible to look for common biographical patterns for such a diverse social group? I will also try to show how moving to another country affects the future of youth, their identity and feeling of strangeness; but, above all, I will focus my attention on the issue of drivers of mobility in various perspectives.

Seeking answers to these questions, I will refer to the biographical research I conducted between 2016 and 2019 in several Southern European cities. Participants in the study came from different migratory experiences and countries of origin. The interviews took place at different times, due to interviewee schedules, but, in total, I collected 70 interviews among social workers, cultural mediators, association representatives, and stakeholders across Spain, Portugal, Italy and Greece. Then I decided to be more focused on the Italian case, which could be identified in the field of migration and in the aftermath of the Mediterranean refugee crisis as an emblematic explanatory case. Interviews were conducted in order to reach a better understanding of the role played by religion in the migratory sphere; starting from personal life, and proceeding with the involvement of religious ethnic players and concluding with faith-based associations and their involvement in the management of both welcoming and integration services. Qualitative interviews were suitable in this case because the main idea here was to understand and get the opinions of the interviewees upon very personal and sensitive topics (i.e. religious and spiritual life, migratory journey, discrimination and difficulties) within the socio-political framework conditioned by the refugee crisis starting in 2015 and still ongoing. Even though the interviews were conducted among members belonging to various religious denominations, this study was not categorized as comparative research because the aim was to understand to what extent religion intervened in the migration field at personal, group, associative and political levels, rather than trying to generate or illustrate theoretical insights through comparison.

The interviews conducted were semi-structured where the questions were prepared in advance including specific questions.

In the book, I refer to various empirical materials:

- a random sample of interviews (60), age range 18–24, equally distributed between male and female.
- 30 interviews with key informants (school managers, local administrators, scholars involved in developing local policies in the various fields of welcoming and societal inclusion: education, training and labour market insertion, housing, ethnic associations).
- a sample of religious leaders or ethnic-religious association representatives in various immigration towns and cities, from Lisbon to Barcelona to Montpellier to Athens. The book is not focussed on their experiences: these interviews have been used as a "check" to verify whether there are similarities among those EU countries where the refugee crisis has been most difficult with the worst consequences for social cohesion.

Much of the conceptual framework I used here—sociological, political, anthropological, network innovation—developed in both an in-depth literature analysis and the process of my fieldwork, and through many joint discussions of my material. Perhaps the most fundamental issue I addressed was the nexus between religion and identity in the framework of migration according to youth perspectives. My process of analysis pushed me increasingly towards this topic, as my field data gradually forced me to understand better the role played by religious attitude in the various migratory aspects, involving migrants, native residents, adults and young people.

Much of the rest of this book describes how religion has become predominant in the last few decades thanks partly to the increasing visibility of migration waves and strongly shaped by local political and social conditions.

Plan of the Book

Ethnicity, Identity and Faith is, firstly, a book about human mobility. Some prefer the word immigration. There are several debates in academic circles on how to label people who cross borders, as well as those who continue to be "foreigners" in the only country they know. Labelling is always a crucial process, which categorizes, stigmatizes and assigns rights or denies opportunities. It is not a novelty. Humanity is a story of migration and mobility. What seems to be exacerbated nowadays is the role played by cultural traits, such as the religious one.

A century ago, sociologists Thomas and Zhanieski pointed out in *The Polish Peasant*, a milestone in migration studies, the role played by religion and ethnic worship places in helping migrants to cope with the challenges in setting up a new life coming from rural Poland to the fast-growing industrialized America.

Faith practices and religion show their role in reinforcing ties to the homeland culture, to serve as a coping strategy with the psychological distress caused by migration, to support and reinforce social identity, especially where discrimination and racism are established daily practices. If there is no doubt that religion is central to immigrants' personal and community identity, at the same time religion is unsurprisingly present in the host societies.

Religion is in the spotlight as never before. In the years that were supposed to witness the triumph of secularization and the advance of the "no religion" category, God (however described) is more than present in the various host societies. Not in the traditional way: churches are almost empty, and mosques and Sikh and Buddhist temples are not in the best of health either. The persistence of religion can be seen in practice (more social welfare than prayer, more advocacy efforts than spiritual), in orientation and in words (welcoming outsiders and the poor, as the Gospels preach), in the gesture of famous people (such as the Pope's trip to Lampedusa, his first outside Rome) and ordinary people (such as the many families who opened their houses to refugee adults and children in Greece, Italy and Spain). The reference to religious values and practices, often invisible and implicit, is a characteristic connoting the relation between immigration societies and those who arrive from elsewhere.

In a broader sense, it also concerns all the citizenry, without forgetting that relational dynamics, in their cultural and economic aspects, are triggered not only between natives and foreigners but also among foreigners of different provenances. Daily contact and competition (for jobs, public resources and services) may lead to intergenerational social fractures and dangerous phenomena of closure, mistrust and exclusion of the weakest social layers.

Migration processes usually involve challenges in inter-generational relations among the universe of migrants composed of families, groups and those who travel alone and, on the other hand, people, social workers, associations and institutions in the immigration countries. Insertion in a host society can have a collateral effect on adult-young and parent-child relations. Indeed, children can interiorize social and cultural values in contrast to what their parents believe or how they behave. Among these cultural challenges, religion plays an important role. Even in earlier migration experiences the focus was devoted to investigating to what extent both first and second generations continue to share the same values (and the same religion).

Chapter 2 inserts the debate on human mobility diversity in Italy into migration studies by providing a general—very brief—reconstruction of the field, outlining three key issues that are at the core of public discussion today: second generations, access to citizenship and the management of religious pluralism. The heart of the chapter is the identity issue and how this topic has become a crucial question in analyses of the intertwining link between religion and migration. Specifically, the focus is on the young, due to their being key actors in the political, academic and daily debates. To what extent will young foreigners or those who have a migratory background be inserted in "our societies"? What kind of activities and initiatives should be addressed to them to avoid the risk of their being radicalized or marginalized?

Adolescent identity definition and social integration sway between a strongly emotional cultural system within the family context and the array of social symbols and meanings conveyed by the host society. For foreign adolescents, the status of "children of immigrants" signifies a greater likelihood of encountering difficulties in accomplishing the full realization of one's own subjectivity as well as a greater probability of marginalization. Like everyone else, foreign youth are called upon to take on a role and work out their identity, which is not merely an individual process but rather involves family, school, cultural and religious environment and friends. Therefore, it is necessary to assess how individual preferences combine with first- and second-generation projects and with integration policies of the host society. Recently, research on children, adolescents and young immigrants has further developed in the direction of studying in greater depth their level of integration outside the school. Ultimately (both inside and outside school), researchers are aware that it is necessary to explore overall social and cultural conditions in order to comprehend the integration and psychological well-being of adolescents and young second generations. More specifically, studies have also explored gender inequalities, identity formation (i.e. self-image, ethnic awareness), home-school-society conflict, inter-ethnic relationships (inside and outside school), attitudes towards religions and value orientations.

Chapter 3 develops one aspect of Chap. 2 further, showing how religion is the crucial arena where both integration and identity paths develop. As presented in Chap. 2, the study of integration processes reaches a crucial stage with the emergence of the second generation of immigrants. Even in a relatively recent migration country, knowledge of the strategic importance of religious environments should be understood as crucial in understanding better young people and providing tools and projects to overcome the risk of antagonistic feelings. Indeed, as one interviewee said,

> If you don't feel at home, why should you have to take care of things, places, legal order? You have to experience how frustrating it is being on a bus and perceiving people are waiting for you to get off, chatting to each other about those migrants who are invading our streets, benefiting from our resources, using our services (M, Senegal, 24, Milan).

The paradigm of structural adaptation postulated a second-generation rejection of parents' cultural identity, including religion. Recent research concerning segmented assimilation and transnationalism suggested that the relationship between the second generation and religious affiliation might differ among diverse immigrant groups (Levitt, 1999). In fact, the religious attitude of the second generation can vacillate between two main positions. On the one hand, religion becomes more of a private matter and its role in other spheres of life diminishes, as the traditional secularization thesis suggests (Billiet et al., 1998). On the other, religion becomes a more central matter, sometimes dealing with the revival of religious ideologies, such as in Islam. The emergence of groups or single individuals fascinated by religious ideologies may be considered as a reaction to the different treatment that Italian society, in which they were born and grew up, dispenses to them: lower-paid jobs, poor housing in suburban areas, discrimination in schooling and the labour market. In this perspective, young people react to unequal treatment and, in many cases, to the challenge of pluralism and secularism in Europe, organizing themselves around a common religious identity, enforcing ethnic identities.

Chapter 4 continues the debate on religion, focussing on Muslims. The chapter questions religious identity among children of immigrants: the issue is on the agenda in many research projects and of governments due to the increasing visibility of Muslims belonging to second generations. In this chapter, I examine the question of whether young people growing up in a Muslim context can be said to maintain their home country religiousness in their new life context or shift to a more secular mode.

Chapter 5 focuses on Catholics, the welcome immigrants compared with Muslims. Religion continues to play a role, an ambivalent role, also when we speak about Catholics and Christians.

Chapter 6 reminds us that host societies are still religious and that this trait should represent an opportunity for the management of migration flows. Indeed, the role of faith-based organizations continues to be crucial in the numerous societal domains of migrants. Cities are the places where these organizations play and show their important role in supporting local administrations. Small or big, part of a broader network or a newly set-up organization strong in religious values, they are key players in the arena of migration, becoming, step-by-step, more important, both locally and transnationally. The case of Waldensians with the Human Corridors project or that of Catholics with the "Rifugiato a casa tua" project, launched by *Caritas Italiana* following Pope Francis' words for the Jubilee of Mercy.

We are coping with a new phase where religious visibility, religious belonging and religious values serving as a life guidance will be reshaping the role of faith both among migrants and natives: this is the challenge facing Italian society, today, and all immigration societies in Southern Europe, as the Conclusions make clear.

References

Levitt, P. (1999). Social Remittances: A Local-Level, Migration-Driven Form of Cultural Diffusion. *International Migration Review, 32*(4), 926–949.

Billiet, J., Laermans, R. & Wilson, B. (eds.) (1998). *Secularization and social integration*. Leuven University Press.

Contents

Chapter 1
Much More Than a Question of Borders

The development and modification of human mobility in recent years have made the issue of borders the protagonist and, consequently, defined the rules for becoming part of the community of those who have rights.

In fact, what seems to have distinguished most the latest period of immigration in Europe has been the change in the phenomenon in terms of arrivals, motivations and permanence on the territory: the differentiation of the demands of immigrants (no longer just those of adults, but also those of families and single children) and the responses of the natives (e.g. in terms of requests for greater control and safety). Old challenges with new outlines need to be faced. The immigrant[1] presence is therefore no longer either a novelty or a temporary exception (see Fig. 1.1). Foreigners have become part of the working environment, of the neighbourhood, of the district. And even more than the adults, their children—important presences in schools, universities, associations and sports. A transversal presence that, beyond the discussions on numbers and entry policies, poses important questions linked to the themes of identity, recognition, and loyalty to one or more countries or places. At the same time, the debate on immigration in Europe has progressively intensified since the years of the economic crisis, with its heavy impact on the employment and income prospects of the population, and taking into account the rapid increase in the number of asylum seekers[2] in recent years (Heath & Richards, 2019).

[1] In the book, we refer to the UN Migration Agency definition: indeed, the term migrant (a) should be understood as covering all cases where the decision to migrate is taken freely by the individual concerned, for reasons of 'personal convenience' and without the intervention of an external compelling factor (Art. 1.1.).

[2] Leaving aside the large categories of labour migration and family reunification, here we recall the scope of so-called forced migration, a definition that encompasses those of refugees and asylum seekers. Intrinsically linked, these terms are not synonymous. Refugees are in a broad sense people who leave their country when forced by external events, thus distinguishing themselves, at least in theory, from those who leave to try to build a better socio-economic condition for themselves and their family. Within this category, the asylum seeker is defined as one who applies for protection, in accordance with international legislation. The refugee status represents the recognition of a legal

R. Ricucci, *Ethnicity, Identity and Faith in the Current Migratory Crisis*,
https://doi.org/10.1007/978-3-030-84056-3_1

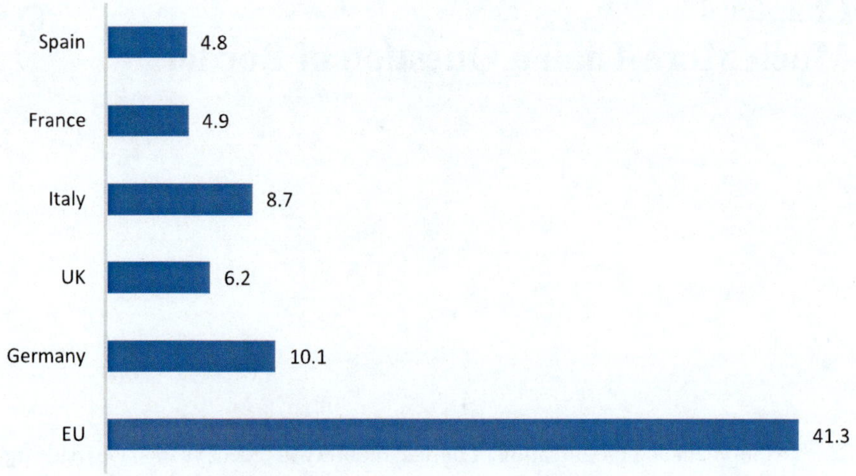

Fig. 1.1 Foreign residents in the EU. Total and first five countries. Data on 1st January 2019. Source: Eurostat database

Although it is necessary not to distract attention from the emergencies and challenges to coexistence represented by the asylum and migration crisis that came to a head in Europe in the summer of 2015, it is equally important to underline the importance (numerical and social) of men and women, families and children who represent another history of migration in everyday life, which populates large and small municipalities, makes multicultural schools and workplaces, and produces fertile contamination.

There are also frictions (Vestergaard, 2020). While these are inevitable and specific to every migration context, recent years have seen an increase in racism and discrimination (Eurobarometer, 2019). Hostility towards skin colour, migration history, professed religion and accent, does not spare anyone. The need for information and training on immigration, cultural diversity, residence rights and mobility vocabulary is very clear, as evidenced by the IPSOS MORI polls (2017). Moreover, many news operators do not contribute to solving the situation, and neither do citizens

status, pursuant to the Geneva Convention of 1951, art. 1 that defines as a refugee a person who: "As a result of events occurring before 1 January 1951 and owing to well-founded fear of being persecuted for reasons of race, religion, nationality, membership of a particular social group or political opinion, is outside the country of his nationality and is unable or, owing to such fear, is unwilling to avail himself of the protection of that country; or who, not having a nationality and being outside the country of his former habitual residence as a result of such events, is unable or, owing to such fear, is unwilling to return to it". This statement has been updated during the subsequent decades accordingly with the historical developments (Chimni, 2009) and nowadays a refugee could be considered "someone who is unable or unwilling to return to their country of origin owing to a well-founded fear of being persecuted for reasons of race, religion, nationality, membership of a particular social group, or political opinion" (www.unhacr.org). It is therefore the last step of a hypothetical path that starts from the need to leave one's home-country to become an asylum seeker up to acquiring a legally stable condition in the country of arrival.

themselves, who populate the fake news network through the megaphone of social networks, reinforcing stereotypes and fuelling fears (Ipsos Global Advisor, 2016; Pew Research Centre, 2016). Today, the construction of news becomes crucial, just as in the debate on immigration, citizenship and access to (scarce) resources (Georgiou & Zaborowski, 2017). Faced with the mass of images and news on the Mediterranean and Eastern European routes, narrations are disappearing dedicated to the (advanced) consolidation of the processes of integration and social cohesion concerning the everyday life of districts, neighbourhood relations, workplaces and leisure associations, schools and recreation-sports, cultural and civic commitment associations. On the other hand, looking at those who have a job or family reunification, who are long-term residents or who are in the anteroom waiting to acquire citizenship of an EU country, is a reminder not to lower one's guard against the needs of those who—even those who are long-term migrants—manifest linguistic, social and economic fragility.

"Easy to say, difficult to put into practice." Even when the scenarios appear similar, the characteristics of the reference populations are comparable. The definitions of policies that have to do with immigrants touch sensitive chords of citizens, or rather of the electorate. Despite the investment in terms of study and analysis, inside and outside academe, the so-called research-policy nexus, the connection between research and political circles appears to be very weak (Scholten et al., 2015).

1.1 Setting the Scene

Migrations are part of world history. The genetic code of populations, from Argentina to Siberia, from New Zealand to Canada, tells of population movement and settlements. Wars, social violence, famine and natural events, the mass exodus or decisions of individuals and families—over time—have designed the atlas of ethnic groups, mixing not only somatic characteristics but also values, cultures, and religions. The processes of mobility change societies, since migrants—as many emblematically repeat—are drops of water that infiltrate the cracks in the ground and then become rivers, until they reach the strength to change everything they find in their path: neighbourhoods, towns and cities.

Nations, in this way over time, have become *de facto* multicultural contexts, multi-religious, multi-ethnic, even if sometimes not recognized *de jure*. Over the centuries, history has shown that it is impossible to stop those who are forced to flee from their own country, or choose to leave home and loved ones to give themselves and their children a future (Castles et al., 2015). Migrants build their inclusion paths crossing all the societal domains, challenging the status quo in each district, school, job sector, cultural environment. As Fig. 1.2 shows, there is no domain which can avoid coping—and interacting—with requests, needs, values, cultural traits and religion.

Migrants, however, not only represent a composite set of backgrounds, biographies, values, pathways of integration but also ownership of rights and policies of integration and citizenship from which they can benefit. Moreover, as far as the

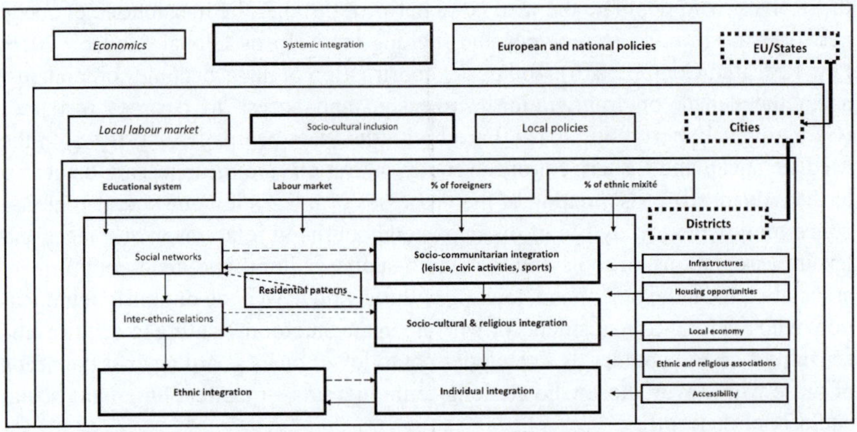

Fig. 1.2 Societal domain-migrants nexus. Source: based literature review on migration studies and discussions emerged within IMISCOE and Metropolis conferences

European Union (EU) is concerned, the mosaic of rules and procedures in the different countries is heterogeneous; in fact, the theme of migration remains a bastion strongly defended by the individual national states against attempts at communitization and harmonization (Bevelander & Petersson, 2014; Estevens, 2018). This situation is emblematically brought to the fore by the recent discussions and decisions concerning the management of the arrival of thousands of refugees and asylum seekers in the countries of the southern shores of the EU and the possible process of burden-sharing in their reception among member countries with, above all, Spain, Greece and Italy interested in a solution as much as possible community—and territorially—dispersed, and the others, from France to Austria to the countries of Central and Eastern Europe on the opposite side (Porumbescu & Pogan, 2018).

Historical memory rarely comes to our aid when discussing migration. The pressure of thousands of asylum seekers on the borders of the common European house is not a new phenomenon. Already with the *Aussiedlers* and *Ubersiedlers* of the 1990s, the newly reunited Germany found itself in difficulty, and alone, in managing such unexpected migration flows (Haug & Sauer, 2007). However, the similarities with ongoing processes end here: in this case, common descent, cultural closeness and shared historical events, together with skin colour and a non-problematic religious belonging, made the news less alarming and silenced anxieties. At least at international level because in Germany the topic was (and is) heated, highlighting the inequalities among the different categories of citizens that exist, among first-class migrants (descendants of German ancestors, although now without any connection to the German language and culture) and second-class ones (children of Turkish immigration, born and raised in the shadow of the economic system and the successes of national manufacturing to which they often contribute as workers, but are hardly recognized as citizens) (Bauböck & Tripkovic, 2017).

Today, the flux of asylum seekers challenges views on diversity and the possibilities of managing socio-cultural-religious pluralism in the cities. Indeed, the arrivals are often from sub-Saharan Africa or Southeast Asia, without a common language (at least with the majority of European citizens, because many speak *Bambarà*, a language widespread in several Central African countries), with religious affiliations ranging from Islam, even in its variant of muridism, to animism to Catholicism. In this context, questions such as "Who are we?" and "Who do we want to become?" come to the fore again. Above all, one wonders how "they" can become "we", or again "Who can we welcome?". In this framework, the religious issues stand out.

Various political and economic causes have contributed to the arrival and stabilization of foreigners in the countries of immigration, forcing nations to set aside ideas and procedures for reception ideally designed for a population that is imagined as temporary and eager to return to the countries of origin or to develop real integration policies (De Bock, 2015).

Considering the traditional European countries that attract migrants (from France to Germany, from Britain to Belgium) and those that have recently become so (Spain and Italy), under the expression *Gastarbeiter* (and increasingly working), adult or minor, there is a set of national groups that live and work with different legal statuses. It is sufficient to think of the distinction among EU citizens, i.e. those belonging to the European Union, and non-EU citizens. Or special historical relations, deriving from the colonial past or bilateral agreements (e.g. in Spain for migrants of Latin American or Equatorial Guinean origin, the period for obtaining citizenship by naturalisation is shorter than that generally provided for other foreign residents). Alternatively, even under legal conditions defined by the reasons for arrival: economic or family reunification, liberation from circuits of exploitation and human trafficking or escape from war (Wihtol de Wenden, 2017).

Therefore, while it is necessary to focus on emergencies, and the needs and challenges of coexistence posed by continuous arrivals, it is equally important to focus on the men and women, families and children, who represent another story of migration in everyday life, which populates large and small agglomerations (OECD, 2018). On the one hand, this attention should be read with a view to describing the consolidation (advanced) of processes of integration and social cohesion that affect the everyday life of neighbourhoods, workplaces and leisure-sports associations, schools or recreational, cultural or civic-commitment associations. On the other hand, a look at those who have a work permit or family reunification—those who are long-term residents or at the antechamber of the acquisition of citizenship of an EU country, is a reminder not to lower our guard with respect to the needs of those who—even with a long period of migration—manifest linguistic, social and economic fragility. Between these two groups, i.e. those who "feel at home and are not perceived as unwanted…" and those who "work, representing the ideal type of foreigner to be sent home…", there are also EU citizens of the "group of 13", i.e. the countries that joined the European Union from 2004 onwards that, depending on the national context, may mean Polish (e.g. Great Britain) and Romanian (as in Italy).

For many of them the feeling of finding themselves "mid-stream" is very much alive, no longer immigrants, not yet recognized and accepted as the other Europeans, such as the citizens of the group of 15, or the states that gave birth to the European Union from the EEC (Graeber, 2016).

The efforts in today's integration follow elements that various research on both sides of the Atlantic has shown: the paths of inclusion can be most heterogeneous and in the passage from one generation to another that process does not always take place, typical of linear assimilation, of progressive abandonment and oblivion of the culture of origin to assume that of the country of residence, symbolically enclosed in a new linguistic, valued, religious identity. Analyses and reports put in place various forces that cannot be traced back to culture alone (according to the cultural perspective there are some "more integrable" cultures than others) or to persistent and immutable inequalities, which confine and allocate immigrants always and only to the lowest positions of the social ladder, to a structuralist perspective (OECD, 2018; Eurobarometer, 2015).

In the game of inclusion—and therefore of positive acceptance—the social class of origin of the first generations counts first. As Adriana Luciano recalls, "…the opportunities for success of the children of peasants and poor immigrants are different, under all skies, from those of the children of middle-class immigrants with a high level of education, even if fathers and mothers had to put their degrees and their qualified professional experiences in the drawer" (Luciano, 2010, p. 14). Family dynamics are important: reuniting with a parent who has created a family in emigration after years of separation can represent a trauma, in addition to that of immigration. Another situation is the experience of joint migration, where all the members, adults and children, are interested in entering as soon as possible (and satisfactorily) in the new life context. Social capital and the realities of the private social context in which one is inserted play a key role. Formed by fellow countrymen or not, the networks in which one is introduced can represent a springboard for insertion or a reference to ethnic identification. Alongside peer groups, there are associations: ethnic, religious or intercultural, environments where one benefits from material and immaterial resources (think of the role of ethnic churches), but often valuable opportunities to share the fatigue of being an immigrant, to receive school help, to be trained in the so-called soft skills, increasingly a pass for job opportunities. The list cannot end without mentioning three other factors: first, the neighbourhood in which you live; then the school you attend; and the (possible) support obtained to quickly fill what are often linguistic gaps or different school programmes, not lack of education. The risk that adults and minors are considered a *tabula rasa* (clean slate) because they do not have Italian vocabulary not only negatively affects the social and work path of individuals but can also result in an under-use of qualified personnel, available only at the cost of language courses for the host country. Finally, the policies of the host country are important: it is not only the rules that regulate entry and control, but rather the interventions towards immigrants, which from regional to local levels set up the infrastructure for integration (Okolski, 2012).

1.2 Foreigners, Always a Challenge for Nation States

Historically, nation states have recognized a set of rights for citizens, but when the post-war reconstruction and the economic boom of the 1960s required immigrant workers to support economic growth, the crux of the latter's access to rights emerged related to citizenship. Therefore, paradoxically, while civil, political and, finally, social rights subsequently became established in chronological order, in the case of immigrants only social rights (and only some of them) connected to the condition of a worker, were attributed as making them "weak citizens", excluded from political participation, from forms of social protection, discriminated against in the labour market, and not recognized at a cultural level (Santagati, 2018, p. 37).

The migration stabilization process brings out new needs and demands of migrants, reflecting on their rights and duties towards host societies, as well as the relationship between citizenship and the full public, political, social and cultural participation of the immigrant, in the political/legal, socio-economic and cultural/religious fields. Therefore, the need arises to reflect on how immigrant status and citizen status intersect, from which the gap between economic and social citizenship emerges, which affects the quality of life of the host society.

The institution of citizenship is the formal instrument through which full ownership of rights and duties is achieved, as well as equality between immigrants and nationals. However, it cannot always be considered an indicator of the level of integration achieved. This is the case of Italy (and also Germany), where an adequately significant number of acquisitions of citizenship does not correspond to a consolidated immigrant presence and an integration process that provides positive signals.

According to the logic of nationalism, all states should be based on nations and all nations should form states: faced with this statement of principle, nationalists have to realize that not only is this principle not fully realized but also as the traditional conceptions of the nation-state and its citizenship have been marked by factors such as the political and economic integration of certain regions, the economic and social role of multinational companies, and movements of international capital. Above all, it was the arrival of immigrants, during the economic growth of the post-war period, and of refugees, during the last twenty years, that dealt a serious blow to this ideal conception.

Various causes, political and economic, have led to the arrival and stabilization of foreign citizens in countries of immigration, forcing countries to set aside the ideas and integration procedures ideally designed for a population imagined to be temporary and willing to return to the countries of origin. If, traditionally, according to the paradigms of the American sociological school of Park and Gordon, integration meant cultural assimilation and acquisition of citizenship, the situation has become complicated since the 1980s due to the size of migratory flows, the reluctance of many immigrants to become naturalized, the unwillingness of some European countries to accept that immigrants have become permanent residents and from the refusal of many emigration countries to abandon rights over their citizens (temporarily) abroad.

Considering only a few European countries, under the expression worker-foreigner there is a set of national groups that live and work with different legal statuses. As we have seen, it is sufficient to think of the distinction between EU citizens, belonging to the European Union, and non-EU citizens. Alternatively, party due to special historical relations, deriving from the colonial past or from bilateral agreements (e.g. in Spain, citizens of Latin American origin or of Equatorial Guinea, the period to obtain citizenship by naturalization is shorter than that generally expected for foreign citizens).

To this must be added a triple challenge launched by the presence of millions of foreigners: that of dual citizenship, in addition to the second generations and asylum seekers. The assumption that everyone should be a citizen of only one country and no one should be a citizen of more than one, is called into question by human mobility and, today, by the possibility of maintaining dual citizenship. In fact, the number of those with the citizenship of two or more countries has increased, due to the increase in the foreign population in immigration countries and to a high proportion of second-generation young people.[3]

Recognizing the incompatibility between the classic model of national citizenship and the stable presence of immigrants, we should move towards a reformulation of citizenship, going from a particularistic concept, based on nationality, towards a more universalistic one, based on a single identity. The normative idea of a citizenship detached from national identity is strengthened, which is opposed to any restrictive or dilatory policy in the granting of the same. The definition of the concept of citizenship and the need for its redefinition are therefore affected by the contradiction between the principle of sovereignty, which is not universalistic, and democratic principles, which tend to be universalistic; a contradiction that is highlighted by problematic relations between "special duties", linked to nationality, and "universal obligations", linked to the dignity of the individual. First, there is the perspective of mutual utility, according to which immigration is closely linked to the economic utility it offers. Then there is the perspective offered by the model of the "division of labour, organized centrally", whereby the social boundaries of a community of law have only the function of regulating the sharing of responsibilities. However, most challenging of all is the point of view that considers the right to emigrate as a fundamental freedom.

The settlement of immigrants has therefore started a reflection on a possible reformulation of the concept of belonging to the modern state. If, at one time, all members of European states were citizens with full legal, civil and political rights,

[3] International agreements were signed to curb this situation, the most important of which was, at European level, the one ratified by the Council of Europe in 1963. The purpose of the Strasbourg Convention, voted on 6 May 1963, was to reduce the number of dual citizens, which, in practice, created problems of legal competence, as in cases of custody of minors born of mixed couples, of military competence or taxation, to which was added the uncertainty of the legal status. However, despite the good intentions, the results were poor. In the 1980s, several countries of emigration and some countries of immigration, including France, proposed a reconsideration of the 1963 Convention, because in an increasingly interdependent and multinational world, it was perhaps impossible to insist on a single citizenship.

today many members of the democracies of the Old Continent are not citizens of the state in which they reside. Considering the complexity of political rights, if the most important are the rights to vote, both active and passive, and to participate in competitions for public offices, the former are the most closely connected to citizenship. Non-citizens are therefore excluded, *ipso iure*, from political life and only through naturalization can foreign citizens from other countries participate in elections. Once again, the axe of citizenship falls on immigrants, creating, however, such an anomalous situation, in the face of which European governments have been forced by internal and international events (civil rights demonstrations, European integration) to broaden the rights of minor groups, including foreigners, undermining the traditional concept of citizenship, now defined as "the last bastion of sovereignty". Immigrants have seen their inevitably lower economic position strengthened (the only exception is the so-called "brain drain", i.e. the movement of professionals, who, if they come from less developed countries, also have difficulty in integrating) by the status attributed to them—foreigners, guest workers, temporary co-citizens—who, in any case, were excluded from full socio-political participation and the possibility of improving their position. The situation in the host societies has improved over time. Considering the labour market, rules have been established so that those foreign workers who have been residing there for a long time, could not be forced to repatriate only for reasons of economic recession or a high unemployment rate, while, looking at the residence status, also in this area there has been an extension of the rights of non-native people. In fact, an increasing number of countries grant a permanent residence permit to a large number of foreigners who have been present on their territory for more than eight or ten years.

1.3 It's a European Issue, Some Say: Not at All, Others Answer

The fields in which the intervention of EC law is stronger and more widespread are those related to visas and border controls, the fight against illegal immigration, returns and readmissions, conditions of entry and residence, and asylum. In particular, with regard to legal immigration, a few legal instruments have been adopted under Art. 63, 3, an EC Treaty, which are currently in force. These establish a set of legal rules concerning a uniform format for residence permits, a single application procedure for residence and work permits, limitations on the admission of third-country nationals, employment and self-employment, conditions of admission of third-country nationals for the purpose of study, student exchange, unremunerated training or voluntary service, entry and residence of highly qualified workers (EU Blue Card), residence permits issued to victims of human trafficking, the status of long-term residents, and family reunification.

The Lisbon Treaty significantly widens the EU's competences for asylum and immigration issues. First, the Treaty sets specific aims of developing "common policies" for both asylum and immigration in the EU, and accordingly empowers it to legislate for "the gradual introduction of an integrated management system for external borders." Second, the Treaty broadens the EU's power to legislate, to define all the rights of third-country immigrants legally resident in one Member State, including their rights to move to and reside in other Member States. Virtually, the only major policy decision left in the hands of Member States is that of deciding how many nationals from non-EU countries they are prepared to admit directly from third countries.

In addition, the Lisbon Treaty removes restrictions on the Amsterdam Treaty to the ECJ's jurisdiction over asylum and immigration policies, which effectively prevented immigration cases from being brought before the ECJ until they had been taken to the highest court in the Member State concerned. Other decisions have been taken[4]. In this context, we have to recall the latest, i.e. the Stockholm programme. It is the third example of such a coherent EU document on migration, after those of Tampere and The Hague. Its introduction highlights migration as a priority issue and states, "The successful integration of legally resident third-country nationals is the key to maximising the benefits of immigration and defines strategic guidelines in this field" (Final declaration, Zaragoza, 15–16 April 2010). Even though relevant international laws and, in particular, EU directives frame integration policies, integration of migrants has always been decided principally at national level. However, notwithstanding the increase of EU competences in the field of immigration and asylum, the core of immigration policies, namely the rules concerning admission of third-country nationals and the concrete rights to which they are entitled in the receiving state, still lies in the hands of national authorities. Given the national framework, we have to focus our attention on both regional and local levels in order to understand how integration policies are managed and how both European and national legislations become practices and contribute to developing social cohesion. The attention at local level does not highlight a linear and regular top-down process. Financial resources represent a crucial point, and this is why the EU funds a lot of projects and initiatives dealing with integration issues.[5]

However, regions of the member states are not directly involved in either the planning or the administration of these funds: a small part (less than 10%) goes to so-called "community actions", and the European Commission itself issues a call for proposals every year—regions can participate, but the selection process lies entirely at EU level—and most of the funds are distributed to responsible authorities in the member states based on a national multi-annual programming strategy.

[4]The Common Basic Principles (CBP) for Immigrant Integration Policy in the European Union adopted by the Council and the Representatives of the Governments of the Member States on 19 November 2004 and European Pact on Immigration and Asylum, adopted by the European Council of 15 and 16 October 2008.

[5]At European level, the funds have been reorganized around three general programmes: (1) Security and the safeguarding of liberties; (2) Fundamental rights and justice; (3) Solidarity and management of migration flows, including the External Borders Fund, the Integration Fund, the Return Fund, and the European Refugee Fund.

Projects need to be divided into groups: (1) Initiatives developed at national level; and (2) Initiatives to be developed at local level.

Various issues have been managed with EU funds, such as those dedicated to: (a) Supporting migrant integration (language courses, cultural mediators, translations of educational materials and definition of leaflets explaining the characteristics of the receiving societies); (b) Improving the integration of migrant children, from attention to unaccompanied minors to emphasis on the school integration of foreign students; and (c) Organizing training courses for civil servants and other persons involved in public services in order to develop intercultural competences.

Focusing on initiatives developed in various EU countries, it is interesting to note that the great majority of them were developed through a partnership among local authorities, schools, NGOs and associations (intercultural and ethnic). This kind of partnership enlightened one of the main characteristics of the European scenario in the migration field: in situations where there was a lack of institutional initiatives, there was great self-promotion of associations and the third sector. In this way, some needs, dealing with the increasing number of migrants, were satisfied just in time. The side effect of this *modus operandi* was the fragility of these kinds of initiatives: they were carried out under annual funding, without any continuity or any final evaluation of the efficacy of the actions.

Recently, some funding provided by local authorities has required a final report on the activity: this was only a self-declaration written by the grantee organization, without any serious evaluation carried out by experts. At national level, there was not a systematic approach to the various goals of integration policies (as defined by documents at EU level). The effect was a development of initiatives and projects at local level, without any co-ordination, with the risk of overlapping and the misuse of both financial and human resources. At EU level, the European Website on Integration was set up in order to supply a global observatory on integration practices developed both at national and local levels in the EU countries.

1.4 From Integration Models to Selection Procedures

To reason on behalf of the integration of immigrant citizens in today's context means venturing into very rugged and slippery terrain. On the one hand, continuous arrivals, images of a human river crossing the Balkans on foot, stories of camps for men and women, sometimes very young, waiting to embark in Libya. News of deaths at sea or among those who try to enter Europe by reaching the Spanish enclaves of Ceuta and Melilla. On the other, attacks and violent episodes committed by young European citizens of foreign origin, often in the name of a vision of Islam and in some cases of social claims. In such a scenario, it is not surprising that slogans are spreading that bring nationalist interests back into vogue, in which ethnic, linguistic and, above all, religious differences do not represent a value or a right to be protected, but rather distinctive traits to be annulled—results of a long wave, symbolically started in the 2000s. Through rules and procedures, tests of knowledge

Table 1.1 Asylum applicants in the EU and in the first five countries (2015–2019)

Country/Year	2015	2016	2017	2018	2019
European Union	1,322,850	1,260,920	712,250	664,410	743,595
France	76,165	84,270	99,330	137,665	151,070
Germany	476,510	745,160	222,565	184,180	165,615
Greece	13,205	51,110	58,650	66,965	77,275
Italy	83,540	122,960	128,850	59,950	43,770
Spain	14,780	15,755	36,610	54,050	117,800

Source: Eurostat

of the language or history, walls are erected that raise the bar not only to be able to look for a job but also to enter one of the so-called countries of immigration: first of all was the Netherlands, with the introduction of a language and civic culture test to be admitted for entry (Penninx et al., 2004). Next, Great Britain, France and Germany (Scholten et al., 2015). Policies are introduced that select citizens who intend to immigrate, or privilege those who have useful qualifications for companies in the national economy, as is the case in Canada. Practices that protect national interests in a scenario that wants, unilaterally, finance, goods and global workers (Table 1.1).

These developments have called into question national integration models and launched a process of in-depth review. In fact, for some time now, scholars have highlighted the fact that traditional models do not respond to the dynamics of migration and the many faces, the varied legal conditions and heterogeneous pathways of integration of foreign citizens, and the strong dyscrasia between the theoretical national framework of reference and implementation at local level (Zincone, 2006). The recognition of this failure in the setting-up of integration policies has given rise to different reactions, from an attempt at corrective policy interventions to the definitive abandonment of the model: one fact emerges clearly, namely the pressure towards a neo-assimilationist model (Caponio & Jones-Correa, 2018). However, it remains to be asked what is the legacy of the so-called models of national integration. Some traces can be found in the civic integration programmes, that is, in the procedures—both selective and formative (of the good citizen)—of verification of the requirements of knowledge of the language and of civic culture to obtain a residence permit, to renew it or to access welfare measures. Therefore, over the years, the tendency has been strengthened, on the one hand, to select, elaborating entry policies aimed at outlining as much as possible the profile of the desired immigrants (generally skilled) who are already literate in the language and culture of the new life context. On the other, the focus is on training the newly-arrived foreign citizen, through pre defined and compulsory paths, whose failure can negatively affect a claim on the possibility of residing regularly in Europe: this is the case, for example, of the Integration Agreement in force in the Italian context. The immigrant, man or woman, adult or minor, to be accepted must become the "good citizen" and—implicitly—assimilate to the new life context.

In the face of migratory phenomena, the universality of rights is necessarily confronted with the safeguarding of cultural specificities; a necessary dialectic is therefore created between the universality and relativity of legal prerogatives whose reciprocal boundaries cannot be defined once and for all. The dichotomy between "assimilationism" and "multiculturalism" comes to the fore, between the tendency to ask migrants to become like "us" and the tendency to recognize "the value of diversity and respect – even in the public arena – for cultural, linguistic and religious traits not shared by the majority".

Emblematic of the difficulty of reconciling respect for human rights, the protection of diversity and security requirements are two hot topics: that of refugees and that of Roma, whether European citizens or foreigners, whose management—transversally to many national contexts—has characteristics of serious problems. In both cases, these are thorny issues, which worry citizens and administrations: the first, aroused by the so-called "entrepreneurs of fear of diversity", and the second, alarmed by requests for government, control and security.

As regards the management of the continuous reception of asylum seekers, from the experience of the last two years we can see the significant commitment often made in concert between institutions and the private sector—social, secular and religious—a "teamwork" that has avoided the occurrence of social clashes and sensational demonstrations of protest. Effective actions against polemics are also the experience of integration and voluntary service.

However, in the face of those who manage to find their way into Spain, Italy or Germany, there are those who continue to write pages of marginality and difficult relationships with citizenship. This is the case of Gypsies, the second enflamed issue on which for years the European Union has been urging (and funding) initiatives to overcome camp policies, promote social and work integration paths of a heterogeneous community, within which European citizens (Romanian or Hungarian Roma, as well as Italian Sinti) and migrants live together, resident groups for generations and others who would never abandon nomadic life. Accommodation in equipped areas or housing units located on the territory to avoid dangerous and rejected concentrations or assisted repatriations: important objectives, which, everywhere and in confirmation of the difficulty of managing the issue, have collided with prejudices and stereotypes that have been strengthened thanks to the viral force of social networks, where refrains of *Auslander raus*", going back to Germany in the 60s, have collected thousands of comments to activate racism and discrimination, and slogans return to the fore that underline priorities and ethnic-identity hierarchies (Ricucci, 2017).

1.5 The Great Illusion: Between Reality and Perceptions

There is no doubt that the migration theme is one of the most important and most controversial aspects of European political debate: in 2016, 48% of citizens considered immigration as the main EU problem compared with 14% in 2010 (EPRS,

2016). Considering an emblematic country such as Italy, according to a survey carried out by ISPI, RAI News and IPSOS (2015), a small minority of Italians (2%) considered immigration as a resource for the country while 67% of respondents thought that migration represented a threat to Italian security, of whom 38% believed that it might even increase the risk of terrorist infiltration. Less than 30% of the same respondents believed that immigration was an inevitable phenomenon that did not directly threaten Italy and should only be at best managed. In addition, indeed, as Table 1.2 shows, the share of foreign-born citizens in EU countries will increase.

A lot has happened in a few years: the severe economic-financial crisis has influenced a reduction in public resources and increased unemployment while ISIS appeared on the international scene and attacks against symbolic places of the Old Continent (from Paris to Nice, Brussels, Berlin and Barcelona) were claimed by the Islamic State. This perception of immigration is certainly motivated also by the significant arrivals of migrants and refugees who have challenged the EU's internal cohesion.

The nexus between these episodes and immigration has been improperly established and exploited throughout the years, particularly by populist, anti-immigrant forces.

In absolute terms, the most numerous collectivities of the more than 35 million foreign citizens live in Germany, the UK, Italy, Spain and France: together, at the beginning of 2015, these five countries accounted for 76.5% of the total. This datum would not be complete, however, without citing the so-called "refugee crisis", or rather, the topic of asylum-seekers trying to remain in Europe.

Focussing on the issue of refugees and asylum seekers (see Sect. 1.6), their growing number, the efforts undertaken by some countries and the resistance of others to their reception, as well as the increase in deaths at sea, have been a real test for

Table 1.2 Share of foreign-born in EU countries' total population: forecast for 2031

	2031		2031
Austria	28.4	Latvia	9.8
Belgium	23.7	Lithuania	8.6
Bulgaria	2.5	Luxembourg	53.5
Cyprus	37.5	Malta	12.7
Czech Republic	11.2	Netherlands	17.3
Denmark	19.2	Poland	1.7
Estonia	11.3	Portugal	18.3
Finland	9.0	Romania	1.0
France	14.4	Slovakia	5.0
Germany	23.0	Slovenia	19.0
Greece	21.2	Spain	29.0
Hungary	7.1	Sweden	23.3
Ireland	29.8	UK	21.8
Italy	17.7	EU total	17.9

Source: King 2015

inter-ethnic and inter-religious relations, as well as for the relationship between inclusive and exclusive society, throughout Europe, but especially in the countries bordering the Mediterranean.

Among these, Italy is the context in which the issue has been most relevant and in which the political repercussions have not been delayed: the last elections revealed a growing preponderance of right-wing and populist parties, attributed by many to the reaction of the population to the growing arrivals and to fear linked to the dynamics of religious opposition: in the public debate all new arrivals belonged to Islam and the fear of being subject to Islamic invasion re-emerged.

Even if the media and politicians in Europe sometimes stress the risk of Islamization and Islamic terrorism, especially after the proclamation of the so-called Islamic State in 2014 and the attacks carried out by ISIS supporters in Europe, it is incorrect to speak of an "Islamic invasion": according to an updated survey at the end of 2017 from the Pew Research Center, there were about 25.7 million Muslims registered in Europe: 4.9% of a population of over 740 million people, with the highest percentage in France (5.7 million, 8.8% of the population) and Germany (about 4.9 million, 5.5% of residents). Between 2010 and 2016, there was a growth of about one percentage point, from 3.8% to 4.9% (from 19.5 million to 25.8 million). The numbers could double by 2050, however, reaching a maximum of 11.2%. In other words, an average of one Islamic EU citizen every 10, just above the current proportion of the French population. Given these data, however, the perception is totally different, as demonstrated by a survey that measured the gap between public perception and reality in 40 countries in 2016 and found that in Italy the actual Muslim population was 3.7% while the perception was 20% (Pew Research Centre, 2017). However, this was not only the case for Italy. The average French estimate was that 31% of the population was Muslim, almost one in three residents, while the real figure was 7.5%. And German and Belgian respondents all guessed that more than a fifth of the resident population was Muslim, while in reality the figure ranges from 5% in Germany to 7% in Belgium.

But religion is not only important as part of the identity of migrants and refugees, or a political argument in the debate, and actions to attribute the so-called refugee crisis a fundamental role, have been and are still being played out by the international, national and local religious organizations in terms of advocacy and concrete projects and strategies.

1.6 When Will We Be Accepted? I'm What My Ethnic and Religious Background Say I Am

When you are an immigrant or a child of immigrants, as in our case, the question seems to hide another much more direct question: Are you innocent or guilty? Are you a threat and a risk or are you attractive and will not bother me? Are you incapable and not seriously committed or are you suitable for what we want from you? (Quote from www.g2secondegenerazioni.it).

European countries have adopted policies for immigrants (the so-called integration models) that differ from each other, which today demonstrate their limitations, their failures, their inability to respond to the needs of an immigrant population, which is increasingly numerous and increasingly interested in becoming an active part, while retaining its own specificities. Even more asylum seekers and refugees want to play this game. Yet there seems to be no room for them.

Always linked to the concept of sovereignty, the status of a citizen is historically understood as the exclusive attribution of rights that are opposed to anyone with a different nationality, according to a statist conception aimed at privileging individual national identities. However, alongside this definition, there has always been a parallel interpretation according to which every citizen is the holder of universal rights. This second concept, defined as a corporate interpretation, identifies citizenship with the participation of the individual in the fate of the community in which they live: in this context, even foreigners can see the substantial differences diminish between their condition and that of the citizens of the place of residence, obtaining from the political powers the recognition of some rights that sanction full integration into the local reality.

The development of integration models (and therefore the decision to modify, or not, the regulations regarding naturalization or granting of citizenship) derives from regulations that have three objectives.

The first objective is that of the "defensive" type, where immigrants are seen as potential competitors on the labour market and in accessing the benefits of social, educational/training, health and housing policies etc., and as potential deviants before which it is appropriate to adopt increasingly restrictive migration policies, such as those adopted during the 1990s by most European countries (Messer et al., 2012).

The second objective is that of "welfare", that is, aimed at combating the marginalization of emigrants who, due to their stigmatization and professional integration, risk being confined to some low-profile niches.

The third objective, defined as "promotional", provides for real citizenship paths, regardless of nationality, for all those who live and contribute to the growth of a given territory and constitute its necessary human capital. According to this perspective, immigrants are offered a real chance of social ascent and full insertion in the arrival society. These three positions have been summarized in literature by defining them as alienship (i.e. those excluded from any right of citizenship on a "defensive" or "custodial" basis), denizenship (i.e. those who enjoy only certain citizenship rights) and citizenship (Spohn & Triandafyllidou, 2003).

In this context, asylum seekers and refugees who, depending on the country in which they find themselves, are confronted with the societies in which they live and what they are part of, who decide what identity they should assume, what faith to profess, which school path to take, where to fit in professionally, and so on. Therefore, even before engaging in the struggle for the acquisition of rights, it seems that this particular category of migrants—different by status and citizenship—share the defence of their identity, the defence of the right to be and to develop biographies by privileging one or two cultural affiliations.

In recent years, naturally variable social conduct has been confused with the identity of a person (plural, but not infinitely) imagined as infinitely variable. If, however, we define identity not as what it appears to be, but as what it is, we are therefore willing not only to defend it from the risks that can undermine it from the outside, but also to build opportunities to rediscover and affirm it. For immigrants all this appears more complex than it is for the natives. In fact, alongside the problems of insertion there are the difficulties of being themselves, of maintaining their identity between "being stigmatized" and "not being considered", between "being placed on the margins" and being considered an expression of a "folkloristic image" that the host society has of their various origins.

In the face of migratory phenomena, the universality of rights must confront the right to the protection of cultural specificities and therefore a necessary dialectic is created between the universality and relativity of rights whose mutual borders cannot be defined conclusively. In this framework the still ongoing refugee crisis puts on the table old challenges. It has been known for some time that the migration issue represents a sensitive issue in European construction: it was the 1980s when discussions began on a possible common migration policy. The fear of a Europe stormed by migrants (especially Muslims) looking for work, jeopardizing the opportunities available to the citizens of the member states themselves, was more than just something in the air. Moreover, still today, one of the recurring fears in the recent debate on refugees at European level refers to the presence of people who migrate for work (so-called economic migrants) among those who request a form of humanitarian protection. This element can be related to the progressive closure of the possibilities of migrating regularly for work reasons: in this specific area, the individual European states have maintained broad autonomy as regards entry policies and the rights to be granted, including the procedure for the acquisition of citizenship.

References

Bauböck, R., & Tripkovic, M. (2017). *The integration of migrants and refugees*. European University Institute.

Bevelander, P., & Petersson, B. (Eds.). (2014). *Crisis and migration: Implications of the Eurozone crisis for perceptions, politics, and policies of migration*. Nordic Academic.

Caponio, T., & Jones-Correa, M. (2018). Theorising migration policy in multilevel states: The multilevel governance perspective. *Journal of Ethnic and Migration Studies, 44*(12), 1995–2010.

Castles, S., de Haas, H., & Miller, M. J. (2015). Walking the tightrope: Between global trends and regional detail. *Ethnic and Racial Studies, 38*(13), 2377–2385.

Chimni, B. S. (2009). The birth of a 'discipline': From refugee to forced migration studies. *Journal of Refugee Studies, 22*(1), 11–29.

Colombo, A. & Magri, P. (2015). *In mezzo al guado Scenari globali e l'Italia Rapporto ISPI 2015*. ISPI.

De Bock, J. D. (2015). Not all the same after all? Superdiversity as a lens for the study of past migrations. *Ethnic and Racial Studies, 38*(4), 583–595.

EPRS. (2016). *Annual activity report*. EU.

Estevens, J. (2018). Migration crisis in the EU: Developing a framework for analysis of national security and defence strategies. *Comparative Migration Studies, 6*, 28.

Eurobarometer. (2015). *Discrimination in the EU in 2015*. European Commission.

Eurobarometer. (2019). *Europeans' opinions about the European Union priorities*. European Commission.

Georgiou, M., & Zaborowski, R. (2017). *Media coverage of the "refugee crisis": A cross-European perspective. council of europe report DG1(2017)03*. Council of Europe.

Graeber, J. (2016). Citizenship in the shadow of the Euro crisis: Explaining changing patterns in naturalisation among intra-EU migrants. *Journal of Ethnic and Migration Studies, 42*(10), 1670–1692.

Haug, S., & Sauer, L. (2007). Aussiedler, Spätaussiedler, Russlanddeutsche: Berufliche, sprachliche und soziale Integration. *Osteuropa, 57*(11), 252–266.

Heath, A., & Richards, L. (2019). How do Europeans differ in their attitudes to immigration? Findings from the European Social Survey 2002/03 – 2016/17. In *OECD social, employment and migration working papers, no. 222*. OECD Publishing.

IPSOS Global Advisor. (2016). *Attitude a l'égard de l'immigration et de la crise des réfugiés dans le monde*. Ipsos France.

IPSOR MORI (2017). *Perils of perception*. https://www.ipsos.com/sites/default/files/ct/news/documents/2018-02/ipsos-mori-perils-of-perception-2017-charts_0.pdf.

Luciano, A. (2010). Introduzione. In R. Ricucci (Ed.), *Italiani a metà* (pp. 9–19). Il Mulino.

Messer, M., Schroeder, R., & Wodak, R. (Eds.). (2012). *Migrations: Interdisciplinary perspectives*. Springer Verlag.

OECD. (2018). *Working together for integration of migrants and refugees*. OECD Publishing.

Okolski, M. (2012). *European immigrations: Trends, structures and policy implications*. AUP.

Penninx, R., Kraal, K., Martiniello, M., & Vertovec, S. (2004). *Citizenship in European cities*. In *Immigrants, local politics and integration policies*. Ashgate.

Pew Research Centre. (2016). *Europeans fear wave of refugees will mean more terrorism, fewer jobs*. Retrieved 1 December, 2020, from www.pewglobal.org/2016/07/11/europeans-not-convinced-growing-diversity-is-a-good-thing-divided-on-what-determines-national-identity

Pew Research Centre. (2017). *Europe's Muslim growing population*. Pew Research Center.

Porumbescu, A., & Pogan, L. (2018). Social change, migration and work-life balance. *Revista de Stiinte Politice - Revue des Sciences Politiques, 60*, 16–26.

Ricucci, R. (2017). *Diversi dall'Islam. Figli dell'immigrazione e altre fedi*. Il Mulino.

Santagati, M. (2018). Turning migration disadvantage into educational advantage. Autobiographies of successful students with an immigrant background. *Revista de Sociología de la Educación (RASE), 11*(2), 315–334.

Scholten, P., Entzinger, H., Penninx, R., & Verbeek, S. (Eds.). (2015). *Integrating immigrants in Europe*. Springer.

Spohn, W., & Triandafyllidou, A. (Eds.). (2003). *Europeanization, national identities and migration changes in boundary constructions between Western and Eastern Europe*. Routledge.

Vestergaard, K. T. (2020). The battle for the public perception: How increases in asylum seekers and media framing affected the perception of the refugee crisis and support for additional measures across the left-right spectrum. *Journal of Contemporary European Studies, 28*(4), 411–433.

Wihtol de Wenden, C. (2017). *La question migratoire au XXI siècle*. Sciences Po.

Zincone, G. (Ed.). (2006). *Familismo legale. Come (non) diventare Italiani*. Laterza.

Chapter 2
Interrupted Routes: Migratory Itineraries and Rhetoric Concerning Difference—News from Italy

Immigration debate in the EU gradually became more exacerbated in the years of the economic crisis and its aftermath. The crisis had severe consequences on the employment and income prospects of the population. Then the rapid rise of the numbers of asylum seekers should be taken into account as another "traumatic" trait of social cohesion and welcoming of migrants in several EU countries.

The Global Compact on Migration and Asylum under the auspices of the UN has shown migration management is undoubtedly one of the main—and most controversial—political challenges facing Europe. In addition to problems showcased in the past few years (such as welcoming and distribution of refugees, managing maritime borders, relations with third countries and legal means of access to Europe), there are others which are less well-known but no less crucial to the economy and social cohesion of the common European area: examples are skilled immigration and the struggle against conditions of segregation, marginalization and ethnic or religious discrimination. These are complicated matters with various implications from points of departure to places of destination, not to mention being interwoven with political, economic, social and cultural vicissitudes informing national demands, capable of stalemating the already laborious process of integrated (impossible in the light of the EU's "federal" structure) management of immigration.

Yet, there has been some tension with regard to sharing and defining common paths, or the moment in the field of principles (such as the 2009 Stockholm Programme of policies concerning justice and internal affairs) or the definition of address documents (such as the Green Paper on an EU approach to managing economic migration, and the subsequent Policy Plan on Legal Migration, COM 669/1); again, in documents dealing with the struggle against clandestine immigration, coordination of migration policies and the necessity to support integration. Finally, there was the definition of a European migration agenda (COM 2015/240) concerning immediate measures face up to the situation in the Mediterranean area, and more general actions to be adopted in the medium to long term. These were attempts to express a European immigration policy, which did not hitherto amount to a true common management of the theme at European level (Geddes & Sholten, 2016).

R. Ricucci, *Ethnicity, Identity and Faith in the Current Migratory Crisis*,
https://doi.org/10.1007/978-3-030-84056-3_2

The evolution of such a complex situation was certainly not helped by some politicians who used migrations as a topic for gaining votes, shouting the slogan "Go back to where you come from! We are closing our borders" (Polakow-Suransky, 2017; Georgi, 2019). Close to various politicians,[1] scholars were invited to pay attention to rising voices of various information operators—as well as many ordinary citizens—who filled the web with fake news, reinforcing stereotypes and fuelling fear by means of social media. Constructing news is today becoming crucial—especially in debates about immigration, citizenship and access to (scarce) resources.

Faced with a multitude of images of the Mediterranean and East European routes, there has been no mention of narrations of consolidated (advanced) insertion processes or of social cohesion in the day-to-day life of neighbourhoods, neighbourly relations, workplaces and leisure-time spaces, schools and sports, and cultural and civic-engagement associationism. On the other hand, a glance at the numbers of those who have had a permit for work or one for family reunion, who have been resident for a long time or are quite ready to acquire the citizenship of a EU country, reminds us not to lower our guard concerning the needs of those who, albeit with long migratory seniority, show signs of linguistic, social and economic fragility. Indeed, statistics on income, level of education, living conditions, and housing segregation, still tell us stories of lack of integration processes (or downward assimilation), stressing the failure of several projects, policies, and good practices developed thanks to exceptional (and not structural) projects. Between these two groups, which is to say those who "feel at home and are not perceived as unwelcome" and those who "struggle, representing the ideal type of foreigners who should be sent home", some strange situations can be seen. First, the state of mind of many citizens of countries which joined the EU from 2004 onwards, is that they are "stuck at a crossroads" between being immigrants and being recognized/accepted as the other Europeans who share the homeland of Adenauer, Schuman and De Gasperi. Then there are the invisible immigrants—housekeepers, caregivers, social-health and nursing personnel—valuable resources often confined to private or care homes (rarely hospitals) where they play a precious role in countries like Italy with one of the highest proportions of the population in the third and fourth ages.

[1] Just to cite a couple of examples of those who gained attention and impact in their national elections, from Marine Le Pen, president of the French right-wing party *Le Fronte Nationale*, to Viktor Orban, Hungarian Prime Minister and leader of the Christian-conservative *Fidesz* (Alliance of Young Democrats), the ruling party in Hungary, to Alexander Gauland, the leader of the German right-wing, populist and Eurosceptic party *Alternative for Germany* (AfD), to Matteo Salvini, Italian Minister of Internal Affairs and the leader of the right-wing, anti-immigration party *Lega Nord* (the Northern League) (Kurecic & Kuhar 2019).

2.1 It Always Happens: Becoming an Immigration Country Means Following Paths Already Taken

Migrations always show the same script and, for this reason, much could be learned by looking back, in the folds of the memory of many, or by looking beyond borders, to try to avoid mistakes already made and instead pursue policies, paths and initiatives that have been able to enhance migrants, especially the younger ones, which each statistic defines as a precious resource for a Southern Europe in strong demographic decline. It may seem a difficult statement to believe in the face of episodes of violence that have these young people as authors: figures full of anger (perhaps hatred) towards societies they despise, considered guilty of not having been able to carry out the proclamations of equal opportunities and equality heralded by the West or with which second generations often grow up. It is not the task of this book to investigate and delve deep into events that have infuriated and dismayed the world. A few years ago, in the face of these episodes, from the riots in London to the fires in Mälmo, to the riots in the French Banlieues, one wondered whether it would also be the case for countries like Italy, which quickly leapt into the top five European countries in terms of the number of foreign residents. Questions were asked if even young foreigners would one day revolt, if they too would feel more North African, Bengali, Chinese (or, even more dangerously, more Muslim) than Italian. The question should rather be, "What can we do to prevent these young people from developing feelings of disaffection and distance from society, so much so that they become foreign bodies and as such indifferent to the fate of the neighbourhoods in which they live, of their peers, of their fellow countrymen?"

In these years, lay and religious associations, schools and local administrations have been active players in the weaving of that invisible web called social cohesion. For some time now, the same young foreigners have joined the work of the loom, committing themselves on two fronts.

The first is that of the universe of the citizens, to show them that they are part of the contexts in which they live, that they are not only physically in Europe with their bodies but also with their heads and that they want to be *pars costruens* of the formative, working, religious and leisure-time realities they go through.

> In many ways we children try to help, showing that we can be Sikhs, young people, cosmopolitans, Europeans. We are committed locally, in our municipality, but also internationally, in transnational youth networks. Yet no matter how much we show that we are young people active with voluntary work, engaged in solidarity associations, it is never enough. You always feel that something is missing. It's not easy, because labels are sewn on you, no matter what you do. I volunteer at a senior centre, but what you see of me is just my ethnic origin, not that I volunteer and help everyone, without thinking about ethnicity, faith, citizenship. (F, 21, Moroccan, Turin)

> I am part of an association dealing with multicultural and multi-religious dialogue here in Barcelona. We organise many meetings, even in the squares, to explain that all religions can live together. We also try to show that parents' migration stories are not repeated in their children's journeys. We young people have a life like all other children of our age. Being children of migrants is just one of the many traits we have. That is why we like to talk and be interviewed, to show who we really are and not let stereotypes speak for us. (M, 24, Senegalese, Barcelona).

The second aspect in which the protagonism of young foreigners intervenes is that of the communities of origin, for which they are important bridge figures, especially for those more fragile subjects who struggle to use a language other than that of origin. At the same time, they represent the hope and concern of the first generations: one is betting on the children to realize that dream of social redemption whose characteristics of the labour market, and regulations and attitudes of citizenship, have made a utopia; the successful integration of the children can, however, result in a progressive departure from the sharing of rites, traditions, customs, marriage choices and religion.

> For some time I volunteered at the mosque in my neighbourhood. I helped the children with their homework. Everything was fine, but when me and the other young volunteers started to make proposals, to ask to be involved in discussions about the activities to be organised, we were 'silenced'. We were told that the organisation and management is the responsibility of the elders. They are afraid that involving us means losing that fidelity to the origins, to the ways of living the prayer and mosque related activities typical of the countries of origin. We are too different now, they tell us. They do not admit that after all they are fond of what is now a faded image: even the countries of origin are changing and even in Morocco, Senegal and Egypt, the way of living religion has changed. (M, 24, Egyptian, Milan).

> In the church where the Peruvian people meet, there are many of us: there are those who organize the songs, those who organize the children, those who entertain the adolescents, those who follow the university students… yet none of us 'children - second generations' have roles of responsibility. We are well known as 'workers', nobody wants to recognize us as 'managers'. They say that we are still young, without experience of responsibility: the truth is that they think we are too Spanish and not very Peruvian. We compare ourselves online with other children of Latinos in Europe: the same experience for everyone. The adults, the first generations, look for us, court us, entrusting us with a few small tasks (always with an elderly supervisor) and then leave us on the sidelines when it comes to deciding. (M, 23, Romanian, Turin).

It is the fear of not having a cultural as well as an emotional bond with the children and, even more so, with the grandchildren; a bond that the acquisition of the same citizenship can break. Symbolically, of course; since it can be translated into an increasingly weaker interest in the country of origin and its socio-political events or, more concretely, into a lack of interest in the events of the national ethnic communities in Italy. However, even for the children of immigration, what Herberg (1955) postulated for European immigration to the USA could come true: "What the second generation rejects, the third generation will rediscover."

Today, parents' fears seem to have been averted in Italy; primarily for two reasons. The first concerns the so-called transnationalism that is increasingly involving second generations: it is no longer just a question of the relationship with relatives in the country of origin in the wake of remittances, but much more, of sharing lifestyles, cultural ties and, why not, work projects based on the chameleonic ability of immigrant children to move among different linguistic, cultural and social contexts. The second involves a citizenship that, even if formally granted, does not guarantee parallel and complete acceptance in society.

The title of Italian citizen then becomes only an element of transition towards full participation in society, which, in order to better explain its effects and counteract the dynamics of discrimination, must be accompanied by educational and training actions for the entire population, from school desks to sports activities, from workplaces to neighbourhood relations.

To date, this is only partially the case, thanks to the work and goodwill of many: the future of the second generations is weighed down by the difficulties of the first generation, of those fathers and mothers who are still struggling to have recognised paths of inclusion. As do general imagination and media attitudes, incapable of (or uninterested in) objectively grasping the wealth of stories and potential, but also of the hard work and problems that accompany the adulthood of immigrant children.

2.2 Us Against Them, Us and Them: The Never-Ending Diversity Debate

Migrants—who pass in and out of the spotlight—have become powerful indicators of citizens' fears and reactions in the face of attacks on their security and their unexpectedly doubtful wellbeing. It is often forgotten that much of that wellbeing is due to citizens who come from elsewhere.

These "useful invaders" (Ambrosini, 1999)—demographic and economic resources as well as cultural consumers and innovators—have been a structural and structuring part of European towns and cities. Our memory has lapsed, accompanied by a degree of myopia, leading us to evaluate our immediate environment without the farsightedness necessary to build—not without debate—processes and paths heralding cultural innovation and enrichment.

Migratory dynamics form part of these processes: arrivals and departures, domestic and international mobility, have always helped—both in the past and in the present—to shape the DNA of societies and populations. More recently, they have done so even in those nations that, for a long time, have been associated with the departure of emigrants and their reverse flow of remittances, memories and nostalgia. This change has transformed territories from which departed first men and then women to destinations that attracted adults and minors from various continents. Many people have to understand this 40-year revolution and then accept it. As for the German Federal Republic in the 1960s, so for Italy and Spain in the 2000s: "We wanted workers ... but we got people instead".

In this framework, Italy stands out for several reasons. Due to its numbers, which have become relevant in a short time-span, and its being at the core of the refugee crisis (see Table 2.1) during 2015–2017 and due to its being a key Southern European Union border.

Indeed, according to the latest available data, at the beginning of 2019 more than 5,000,000 foreign citizens lived in Italy, 8.3% of the total population. In 1992—a key year that witnessed the promulgation of the current citizenship law—there were 356,000, amounting to 0.6% of the total.

Table 2.1 Foreign population: total of minors and their rate of the overall population (2006–2019), data as of first January 2019

Foreign population	2006	2013	2019
Total (adults + minors)	2,670,514	4,387,721	5,255,503
Minors	665,625	953,785	1,082,634
% of minors on total population	22.6	21.7	20.6

Source: ISTAT

Yet, the data are clear in defining Italy as a culturally and religiously plural country: in 2020, its large and smaller towns and cities, and countryside, can be described in terms of the great variety of origins, languages and relations with the sacred, which are linked neither with Catholicism nor with the historical presence of Waldensians and Jews.

To be more precise, this is not a recent phenomenon. Already at the start of the new millennium, the state was no longer homogeneous: over four decades of migrations has profoundly transformed its social and cultural fabric. At least this is the x-ray produced by mapping presences, territorial distribution, and attendance of non-Italian pupils in school classes and observing neighbourhoods whose commercial texture has been designed by ethnic shops and immigrant businesses. On the other hand, collective imagery seems to be mainly aware of emotions and reactions resulting from phenomena that are more striking from a communication point of view and frequently problematic cohabitation. This explains why cultural diversity is presented solely in terms of disembarkation from rubber dinghies, welcoming difficulties and the faces—but not the background biographies—of those who are struggling to rebuild their lives in Italy.

They are men and women through whom Italian multiculturalism can be read, leading actors in the increasing diversity of skin colour, accents, sounds and smells very different from those considered traditional. Chinese, Indian, Moroccan and Latin American food is now part of the diet of the young (and not so young): today, as happened with internal migration, integration passes through cuisine (Bonifazi, 2013). For those who love sport, it is by now obvious that in competitions of every level, foreign (origin) athletes take part side-by-side with the autochthonous. In addition, this without even considering workplaces, commercial and service (usually personal) activities, neighbourhoods and close-up relations. Yet, despite a multiplicity of daily multicultural experiences, memory is selective, as in few other contexts, such as in that of encountering otherness. Indeed, there is a year-by-year increase in signs that the daily lives of Italians are gradually becoming more interwoven with the vicissitudes of citizens from other countries. In the worlds of work and home, in shared urban spaces (from markets to public transport) and in cultural and political debate, everybody can recognize the importance of the presence of immigrants; and their children, the so-called next generation who are also referred to as "new Italians". This cohabitation has various facets: while for some it is a *fait accompli*, and an opportunity for many significant integration experiences, for others it is overpowering evidence of the limits of Italian society and is often used to stir up fear of invasion and an unbearable burden on the country's resources. Being de facto a multicultural country does not mean that the awareness of a such important change is widespread among the citizens. Several efforts have been made through projects in schools, where multiculturalism is evident (around 10% of the students in the compulsory educational tracks have foreign citizenship; MIUR 2020) and children of immigrants grow up close to Italian peers. However, other issues still continue to enforce the debate on how to reduce the migrants' number in the country, such as—among others- the relationship with Muslims, the idea of introducing selective migratory policies and the debate on citizenship (Ambrosini, 2013; Testa & Armstrong, 2012).

2.3 Not Just Being but Also Becoming a Multicultural Country. The Role of Schools

Since schools are certainly the privileged observatory for detecting a country's demographic and socio-cultural changes, they are the most studied context. Data about pupils with non-Italian citizenship indicate a growing number born in Italy and therefore of the second generation. At the same time, they show that for some years the greatest incidence of second generations has been in pre-schools and the early years of primary schools, to diminish later in the fifth year of secondary schools (MIUR—Ministry of Education, University and Research, various years). This tendency is in line with the characteristics of a relatively recent immigration country where most of the foreign population is still made up of first-generation and family-reunion immigrants. Research findings have also demonstrated that in Italy teachers and school managers are dealing with the classic themes of every immigration country: pupils arriving at various times during the school year; the debate about whether to insert immediately the new arrivals in normal classes or to create "ad-hoc classes" (welcoming classes or—as they have been termed during a heated public debate—"bridging classes"); intercultural education; orientation; the choice of educational paths apart from compulsory schooling; and relations with non-autochthonous families and their involvement in an academic system which is often very different from the one they have left behind in their homeland.

These are crucial, strategic questions for the success of any educational stream (even more in an immigration context) which for years have engaged teachers, managers and private-social volunteers. The growing number of second generations has appeared as a soothing factor: the adagio—not made explicit—that considers teaching them as a return to less tedious work: education without language-comprehension problems, the necessity of cultural mediators or producing different paths and programmes.

The advancement of pupils born in Italy is greeted in schools with a sigh of relief. Finding themselves faced with pupils born in the same country as the others, or at least ready to leave the starting blocks with some experience of socialization in an Italophone environment, reassures workers in the field. However, for children of immigration being born in Italy, attending pre-school and arriving at primary school without having lived in another country does not mean being possessed of a lexical store and syntactic-grammatical skills in Italian equivalent to those who have had only the Italian context as a point of reference. Often for second generations, time not spent in school but with the family, relations, in non-organized playtime, means being immersed in a different (and considered less prestigious) so-called "linguistic pool" from the school's institutional environment. Therefore, one cannot rest on one's oars even with the second generation, assuming that the time for personal language intervention has passed. The fact is that this category has specific characteristics and needs linked to various factors: (a) relations with peers, often negatively conditioned by a "different" physical appearance or by one's socio-economic level of provenance (less access to consumer goods or a different family typology); (b) relations with parents who, in addition to belonging to another generation, originated in different cultural contexts from that in which their child/children is/are growing up and (c) environmental relations regarding their origins and mother tongue.

If they are not helped at the right time, there is a risk that the survival strategy adopted by the children of immigration should be that of assimilating the Italian model, with the attendant obliteration of the wealth of their identity. Those who carry out educational roles would do well to reflect on—in addition to what has been already said—further elements related to the fact that many pupils born in Italy are not Italian citizens (Ricucci, 2020). If in an initial (superficial) analysis, we may applaud teachers who do not stress the difference in everyday class relations.

Further reflection reveals that—precisely in educational tasks—young people are being trained who are able to move freely in society, to grasp tools and methods for understanding who, and where, they are, acting suitably in different social contexts. For some time, attention to individualised educational plans—capable of taking into account the specifics of each student—has been developed. Personal and family biographies, the mother tongue and socio-cultural references transmitted by parents are all factors contributing to the student's identity, which the school must face up to and interact with (beginning increasingly from pre-school). Care must be taken with the initial phase of education in a scenario of second-generation stabilisation and growth.

Little research has focussed on how foreign families have made use of pre-school services, but that little amount has pointed out the strong and weak points of a relationship which is still coming into being. On the one hand, foreign families should be convinced (contrary to the cliché that their children are stealing school places from Italians) of what an opportunity pre-school is. They should also be helped to realize how useful socialisation activities, not explicitly aimed at academic learning, are for the education of all minors—a good preparation for subsequent school insertion rather than a waste of time. On the other, Italian families need to be worked on. They are divided between the timid (for whom even pre-school in the presence of non-Italian pupils could have a negative impact on their children's education) and those already aware that from pre-school onwards tomorrow's society is being tested; thus the *mixité* of classes offers a unique educational opportunity to approach what has for some time been a multicultural reality.

2.4 Integration Paths Through Girls' Perspectives

The female protagonism as the leader of migratory chains is combined with the role—sometimes played in the shadows—of women who arrive in Italy to reunite with their families and who sometimes do not succeed (or do not want to) in interweaving relationships with the surrounding environment. Women and girls of foreign origin can thus remain relegated to family communities or the larger ones of the national, ethnic or religious community. Little is known about such situations; however, often service workers, teachers, and private social workers reveal how fragile their condition is within the home. For the daughters of immigration, origin can become a heavy burden and somatic features or linguistic interference can be stigmatising and, in these cases, the label of "foreigner" risks being heavier than that of "young". Above all, it risks overshadowing the countless facets of being young people of foreign origin. Yet, among the different paths and possible definitions of identity, some common traits can be grasped, without being included in stereotyped images. First, plurality and flexibility seem to be two key words to describe the trajectories that young female immigrants, or those of foreign origin, are following. Plurality because there can be many experiences and identity strategies assumed. Also because the individual, family and context conditions (from the neighbourhood to the municipality) in which one is inserted are very heterogeneous; flexibility, because they demonstrate their extraordinary ability to move among different areas, sometimes giving rise to innovative intercultural experimentation. In this way, they do not distance themselves from many of their Italian peers, who build their biographies in an increasingly cosmopolitan and less standardized way than the generation of their mothers and fathers. Both groups, in fact, through daily experiences and encounters/clashes with the many diversities of today's society, give life to the mosaic of their biographies within societies that are experiencing the turbulence of an economic recession and its effects on delicate processes of social cohesion.

The growth of second generations also in this case presents new aspects; in particular, women, who are the protagonists of a restricted sociality, within the extended family nucleus, are highlighted in other areas of socialization.

Interesting findings from this point of view concern the world of sport, indicative also for its strong symbolic importance. The history of immigration in Italy is full of episodes in this regard; for example, the story of Samia Yusuf Omar, a Sudanese Olympic athlete who participated in the Beijing Olympics in 2008, drowned off Lampedusa in 2012 while trying to enter Europe, caused great emotion. However, it is in everyday life that the link among sports of immigrant children has slowly become very strong, particularly through the participation of young foreigners in youth teams. On the one hand, sports activities are indicated as an excellent tool for socialisation; on the other hand, the results obtained by many athletes of foreign origin are seen as a possible element of contrast to discrimination and discussion in positive terms on immigration. Thus, the national records recorded by Ayomide Temilade Folorunso, an Italian hurdler of Nigerian origin, or the victories in rowing of Khadija Alajdi El Idrissi of Turin, daughter of the imam of a mosque, can become symbols of successful integration in national newspapers.

Another important example refers to the presence in ethnic associations, which is becoming stronger, not only in the field of religious commitment and the provision of services to the community, but also in projects to promote integration and social inclusion. After a season full of analysis of the associative reality of migrants, we are witnessing a phase of less attention: there is a lack of an updated look by researchers exploring their development (or implosion) methods, their intervention areas, as well as the comparison (and sometimes a real clash) among the generations of mothers and daughters on objectives, management methods, and relations with local administrations.

As already seen, also in this case the next step is represented by politics. In this context, we can see some (weak) signs of greater civic participation, which are also important from the point of view of their identity. For example, the presence of women is significant in cultural struggles for the reform of the right of citizenship. It should be remembered here that some young people, who have become Italian, first and second generations, are now members of City Councils, directly involved in the management of public affairs at local level.

While it is true that the right to vote is the highest form of exercising citizenship, it must still be said that women are numerous and visible in their commitment as active citizens, especially through their social role. What then becomes central is the heterogeneous, and at times tiring, process of inclusion of young people of foreign origin as cultural mediators, trade unionists, members of employer organisations or volunteers in inter-ethnic associations.

2.5 Migrations and Islam: Still an Open Sore

In 2009, public praying of many Muslims in Cathedral Square, Milan, caused a sensation. Various voices considered this manifestation as an outrage, interpreting it as a threat to a symbol of Catholicism and an iconic site of Italian identity. In this and other gatherings, a request was made for respect for the right to have one's own place of worship. The physical protests were accompanied by a web campaign showing Muslims praying in squares, houses, suburban spaces, open areas granted by local authorities—as long as they were far from the city centre.

Some questions arise from this parade of images. First, the absence of women. No matter how far this fact is from public awareness, the Muslim world in Italy is by now gender-balanced: women form an integral, participating part of prayer ceremonies and within associationism. Therefore, we should not mistake women's absence from leading roles on the public stage for a condition of invisibility. Various research has highlighted that women—especially young Muslim women—are active participants and in many ways represent a powerful impulse in community life.

While the women are absent from the communicative point of view, there is no lack of reference to immigration: in the media and in majority opinion there is no sign of the Islam-foreigner binomial fading. Who knows whether in the near future this perception will start to crumble in the face of Muslims with Italian citizenship, perhaps children or grandchildren of immigrants, who do not even have the migratory experience of moving from the country to the town and city. Undoubtedly, international events and the refugee crisis (which is not however composed of a Muslim majority) provide an alibi for those who forget that Islam, its history and relations with Italy have deep, ancient roots. The people of Genoa know this: during prayers at the end of Ramadan, in 2018, many remembered that at the time of "La Superba", when Genoa was a maritime republic, the city had a mosque with a minaret competing with bell-towers and other buildings. Moreover, this mosque was in *Darsena* (the inner harbour of the city), an area where the *Eid* prayers were held. Ruins of the temple building are incorporated in *Palazzo di Scio*, headquarters of the University's Department of Business Studies.

It is not only a question of the past, important as a part of memory and identity, or weak knowledge of the history of religions. The subject belongs to the present and its identities, which may be numerous and multiple even among citizens of the same country. There is no doubt that immigration has helped to place the theme of religious pluralism at the forefront of debate, but it is important to remember that the theme pre-dates the 1980s migratory flows that turned Italy into a multicultural country. Long before the arrival of the immigrant, there were Italian Waldensians, Protestants, Jews, Jehovah's Witnesses, Buddhists, Orthodox Christians and Muslims. However, by their presence the immigrants made an essential contribution to highlighting a vital question for the life of Italian cities, a subject that, since the 1990s, existed only in ministerial documents or in those relative to the Catholic Church's interreligious dialogue.

While many religions have cohabited over time in the country, their followers—from North to South—declined differently according to traditions, customs and family experiences; there is still widespread conviction that immigration is the only lens through which one can understand religious pluralism. The (well-founded) fear is that of a process of reductionism and simplification of Islam on the part of mass media, online forums and in everyday life situations.

2.6 The (Ambivalent) Desire for a Selective Migration Policy

Over four decades, not only the numbers have undergone change but also the faces and characteristics of men and women, adults and minors, resident EU citizens (more than a million and a half, let us not forget) and non-EU.

In the new millennium, the non-Italian population is no longer a temporary aberration but a stable component of society, often entering the national community by obtaining citizenship, with important repercussions on the demographic, scholastic, employment and—more generally—socio-cultural levels. Before continuing, however, it is necessary to make a premise: life-paths, biographies, struggles and exploits are hidden behind these data, which make the variegated set of ethnic communities even more multi-faceted. Within them, further particularities mark out the experience of minors reunited with their families and outright second generations.

If, therefore, the immigrant presence is neither a novelty nor a temporary aberration, it comes as no surprise how little is known about it and how much it is feared.

Whereas the meeting between natives and migrants may be seen as an extraordinary opportunity for reciprocal enrichment, there is no hiding its tension, above all when it takes place in territorial settings where social cohesion is problematic. In these situations, cohabitation has witnessed intergenerational fissures causing dangerous phenomena of closure, mistrust and exclusion of the weakest social segments—aggravated by the reduction of public resources and the tendency towards welfare-state crises.

At the same time, the gradual stabilizing of foreigners and the rise of entrepreneurial activities connoted as ethnic and interethnic (aimed at a mixed clientele) are changing the nature of neighbourhoods. Nor is this settlement free from difficulties; for example, it has to face the mobilization—supported and fuelled by the media—of elderly residents who rediscover a local identity and belonging and advance claims of protection and security against the arrival of "the foreigner" or, to keep it simple, "the criminal".

Italians still consider immigration their main problem. Paradoxically, just when the numbers of the phenomenon have stabilized after years of growth, the foreigner is a source of worry. Mainly the men because they occasion fear while the women are tolerated in that they are functional to the private welfare assistance mechanism managed by families. The main indicators of society's perception of the theme show information contaminated as to numbers, faces and conditions: everything is blamed on the emergency and the rest is forgotten. Thus, it comes as no surprise that a recent survey revealed that 42.9% of Italians consider non-Italian citizens as a threat (Garelli, 2020[2]). They are competitors on the job market and for social assistance, but—as we have already seen—they are carriers of irreconcilable cultural and religious values. Once again, we could remind ourselves that many, when answering, probably did not stop to consider how many Bulgarian and Romanian citizens, as well as French, German and Spanish, cannot be either denied entry or sent back. Alternatively, they forgot the 22% of immigrants employed in personal caregiving, dwelling as they did on problematic aspects capable of catalyzing attention, casting a shadow over robust, widespread integration processes. On the contrary, 32% of interviewees hold that immigrants are a resource for the country. An identikit of the holders of the two opinions shows contrasting profiles: those who feel threatened are men rather than women (46% vs. 39%), have low educational qualifications, and the elderly; those who consider immigration a resource and an opportunity for the country tend to be young, well educated, and women.

[2] These findings come from a survey carried out on a representative sample of the Italian population in 2018.

Two pieces of data may surprise those not engaged in the field. Among the various geographical zones, it is the North-East which most favours immigration, at 38%, confirming that (despite anti-foreigner slogans and a strong right-wing party leading the great majority of towns and cities and the Veneto regional government) in an economically advanced zone, the positive effects of immigrant manpower—for some time highlighted as a precious, well-integrated resource—on the productive fabric are considered pragmatically (IDOS-UNAR, 2020; Richardson & Colombo, 2013). Another curious point is the anti-immigration stance of many (extreme) left-wing electors. This is in line with what Masci for the Pew Research Centre revealed as far back as 2017, confirming a shift of values on the part of those close to positions defined as "democratic" or "progressive".

These are striking elements in a country which could draw from its own history as an emigrant population, to describe how, through the migratory process, economic, cultural and social resources can contribute to ascending social-mobility paths. Today, as yesterday, therefore, the relationship with otherness is an unresolved question. Debates and data, political positions and legal measures that attempt to exclude, limit and contain it are national examples.

Therefore, immigration is increasingly a fundamental element of politics and public communication, which are at the same time creators and witnesses of the growing importance of positions contrary to a multicultural perspective or to immigrants themselves, seen as dangerous competitors for precious resources (work, welfare services and so on).

2.7 Italy for Italians by Origin. A Citizenship Law Out of Time

In 1992, when the current Law (No. 91) concerning the acquisition of Italian citizenship came into force, immigration was already a well-known phenomenon: not long before, an important legal amnesty entered into force and a Law (No. 939/90) had initiated the first attempt to regulate the presence of foreigners, unconsciously opening the way to what would become a normal praxis in managing the phenomenon. Above all, minors were still few and little visible, whether reunited with their families or born in Italy. On the contrary, the memory of a nation of emigrants, and the necessity of maintaining contacts with Italians abroad and their descendants was still fresh.

As we have seen, in 2020, non-Italian citizens are a structural element of the peninsula's productive and social fabric. The children of immigration are reinforcing the youth age ranges, giving rise to a demographic crisis with serious consequences where the birth rate is lower than that of substitution. Nevertheless, it is a society in which citizenship is still prevalently determined by *ius sanguinis*.

According to Art. 1, Paragraph 1, of Law No. 91/1992: "a) the child of a father or a mother with Italian citizenship; b) anybody born in the Republic whose parents are unknown or stateless, which is to say, if the child is not a citizen of the country of origin of the parents, according to the law of that state is an Italian citizen." Today many claim that the granting of citizenship based on blood rather than place of birth, socialization and education, is an idea that has had its day (Zanfrini, 2020). It is a thorny subject because talking about citizenship touches upon identity, recognition, rights, enfranchisement: in reality, it means imposing a border between those inside and those outside the social pact, access to rights and the possibility of benefitting from resources (Ricucci, 2015a, b).

The question has been, for some time, present in political debate, as evidenced by the numerous bills presented by various political factions. They have, however, three elements in common:

a. Maintaining the *ius sanguinis* in specific cases;
b. Introducing a moderate form of *ius soli*, transformed in time into *ius culturae;*
c. No provision for a radical step from *ius sanguinis* to *ius soli.*

On this subject, political debate is particularly heated, as can be seen from the fact that none of the proposed bills has become law. Often they drag on from one legislature to another. The 1992 law was a product of the lingering image of a migrant Italy, seeing that it fast-tracked and facilitated citizenship for descendants of emigrants. Thus, in countertendency to the wind that was then beginning to blow in Europe, Italy defined itself as an ethnic-based nation made up principally of the children and descendants of Italians. Zincone described this position as "legal family spirit" where joining the "family of citizens" came about by means of blood, descent or marriage (Zincone, 2006).

As we have hitherto attempted to demonstrate, interest in the subject has grown because of the profound transformation effected by migratory flows running through Italian society in various ways. Furthermore, numerous sensitization initiatives have brought to the fore the activism and leading role of many groups and associations consisting mainly of children of immigration.

The phenomenon is still numerically contained, but has a significant impact on specific local realities (important urban neighbourhoods, medium-to-small communes) and even more because of its symbolic value since its promoters (non-Italian citizens) have no right to vote actively or passively. Internet, compatibly with protagonists' ages, is the main tool: the web is essential not only to put forward one's point of view but also as an instrument of updating and information. The latter becomes even more crucial in that—as already seen—the common perception of the migratory presence in Italy is based on inexact data and incomplete information, accompanied by little knowledge of the regulations on the subject.

In virtual debates, on means of information, in everybody's daily life, people get hot under the collar because the subject is connected with rights and duties, which are very little theoretical and are a part of everybody's experience. And, naturally, it is linked to the broader debate about immigration and to what extent a community is disposed to welcome first of all residents and later future citizens (Ambrosini, 2010).

Because of their complexity, these aspects constitute a heated element of ideological conflict, illustrated by the slow, difficult institutional procedure towards modification of the citizenship rules—despite numerous reform proposals– it has not yet produced any results. And in the meantime, children of immigrants enter their adulthood.

Foreign youths go to work and start a family earlier than their Italian peers who increasingly delay these stages, considered as a passage from "youth" to "adulthood". In this sense, the former represents a novelty in the Mediterranean model of adulthood which characterizes Italy. This awareness is not yet widespread: second generations are discussed above all regarding the reality of students, primary and secondary schools; but there is less talk and reflection about universities, where there is an increase in the enrolment of non-Italian students with Italian high-school diplomas. For various reasons, there has been little research into this recent development. Indeed, the numbers are not yet significant and the spotlight is on students coming from abroad, examples of a process of internationalisation, which interests Italian universities. Furthermore, it is true that tensions (from welcoming to didactics, from relations with families to intra-class dynamics) characterize schools at the compulsory level. On the other hand, common perception seems to crystallise children of immigration as being children or adolescents.

Therefore, the common perception seems destined to condemn second generations to be linked perpetually to their family background. They are rarely considered just as young people. Yet, they form part of that composite youth reality, about to become adults in an age of great fear and little hope. Among the concerns of young foreigners are those connected with juridical matters.

Being born in the country does not entitle one to Italian citizenship. Nor does it protect one from processes of stereotyping that even successful study and professional paths fail to cancel in the minds of those who consider themselves the only authentic guardians of residence and governance.

Faced with these attitudes, second generations are beginning to speak up: they get involved in associationism; they take part in debates about citizenship and immigration; they promote sensitization initiatives; they present and debate their lives as "non-citizens" in the only country they really know, where they "feel at home" because they speak the language, have done all their schooling there—and now, for many, university. Through various civic-engagement initiatives, this generation of hyphenated (Italo-Moroccan, Italo-Chinese and so on) future Italians has come to the fore in a new guise—they no longer need Italian lessons as did those who were divided between the needs of the family and those of society. It is the generation that makes its voice heard, writing and programming digital products; which promotes knowledge and information initiatives about what it means to be children of twentieth-century human mobility, and which, making use of online communication tools (from websites to Facebook pages, Instagram and TikTok) to broadcast their perspective on the implications of being non-citizens in a country they consider their own. Through a variety of forms, and channels, and in collaboration with private-social and local-institution subjects, they have taken shape as a social group distinct from their parents (and from newly-arrived migrants) and given proof of their capacity to become important interlocutors for the construction of social cohesion.

The time has come to take a further step—to enter politics. For some it is already a reality, for others a possibility and for yet even others, (still) a dream accompanying that of becoming citizens. Naturally, not all of them will become politicians, but the possibility of choosing represents a further step on the road to feeling properly part of the only life-context they know.

References

Ambrosini, M. (1999). *Utili invasori. L'inserimento degli immigrati nel mercato del lavoro italiano*. Franco Angeli.
Ambrosini, M. (2010). *Richiesti e respinti. L'immigrazione in Italia. Come e perché*. Il Saggiatore.
Ambrosini, M. (2013). Immigration in Italy: Between economic acceptance and political rejection. *Journal of International Migration and Integration, 14*(1), 175–194.
Bonifazi, C. (2013). *L'Italia delle migrazioni*. Il Mulino.
Garelli, F. (2020). *Gente di poca fede*. Il Mulino.
Geddes, A., & Sholten, P. (2016). *The politics of migration and immigration in Europe*. Sage.
Georgi, F. (2019). The role of racism in the European "Migration crisis": A historical materialist perspective. In V. Satgar (Ed.), *Racism after apartheid: Challenges for Marxism and anti-racism* (pp. 96–117). Wits University Press.
IDOS-UNAR. (2020). *Dossier statistico immigrazione. rapporto 2020*. IDOS.
Herberg, W. (1955). *Prostestant-Catholic-Jews*. An essay in American Religious Sociology. University of Chicago Press.
MIUR (2020). *Allievi con cittadinana non italiana*. MIUR.
Polakow-Suransky, S. (2017). *Go back to where you come from. The backlash against immigration and the fate of Western democracy*. Hurst and Company.
Richardson, J. E., & Colombo, M. (2013). Discourse and politics of migration in Italy. *Journal of Language and Politics, 12*(2), 180–220.
Ricucci, R. (2015a). *Religious education in the Facebook era in the Moroccan Diaspora: Muslims on line, young people off line* in Paper presented at the ECER Conference, Budapest.
Ricucci, R. (2015b). *Cittadini senza cittadinanza. Immigrati, seconde e altre generazioni: Pratiche quotidiane tra inclusione ed estraneità*. Seb27.
Ricucci, R. (2020). Discutere di pluralismo religioso in un Paese multiculturale. *Vita e Pensiero, 1*, 52–57.
Testa, A., & Armstrong, G. (2012). *"We are against Islam!": The Lega Nord and the Islamic folk devil* (pp. 1–14). *SAGE Open*.
Zanfrini, L. (Ed.). (2020). *Migrants and religion: Paths, issues and lenses*. Brill.
Zincone, G. (Ed.). (2006). *Familismo legale. Come (non) diventare Italiani*. Laterza.

Chapter 3
Religion: A Neglected Variable in the Migratory Routes

> On Sundays one dressed up and went to Mass. It was a time when one could be oneself, feel at home, hear one's language, the sounds one recognized. It was our time, Italian time. When I tell my grandchildren about it, I feel old. In the church in that period, we spoke a mixture of English, Italian and Sicilian. It was our way of understanding one another. Today, it would be impossible: it's hard enough to get our grandchildren to go to the church (F, 64, Italo-Australian, Perth).

> Our parents don't understand. They still think it's a whim of ours not to want to go to the mosque. They don't understand that we want to be something else. We are Muslims, but in our own way. They live in another age. Many parents, even after years, still don't realize that they are living here in Germany, not in Turkey or Morocco (26, Afhgani, Berlin).

> Here, it is not like France or Belgium: if you are a foreigner here, you are second class. We have to think about learning trust and respect, because this is our country. Religion doesn't help any of us. Even my Catholic friends who have the misfortune of being Filipino or Peruvian don't find it easy—and we live in the Pope's country! Consider that they discriminate against us Muslims for the usual reasons: terrorism, fear, stereotypes. Then you observe that even the others—Catholics, Sikhs, Orthodox, Protestants—are treated no better, so you start thinking that it's not the religion that bothers them: it is us, our very presence, that bothers them: it is us, our very presence, which is causing them disturbance. When you realize that, the thought grows that a lifetime will not be enough to be accepted. Maybe my father is right in wanting to live and to be—including praying—as if he were still in his homeland. Why make so much effort to change if you're always going to be a foreigner? (F, 26, Filipina, Milan).

As always, certain characteristics of migration history are repetitive. The times and the scenarios change: Australia in the 1950s was not the same as the cosmopolitan Berlin of today; and Milan, vaunted as a "super diverse" city, has a much more recent relationship with Islam than the German capital. As always, migrants are asked to enter on tiptoe or wait at the window for some time. Moreover, from the window (a metaphor often applied to associationism and ethnic religious places) one looks at the surrounding world; from a safe, welcoming, reassuring space for everybody, migrants and the welcoming society. It is a space which guarantees reciprocal indifference during a non-active period. As if to remind us that the *Gastarbeiter* image (useful on the labour market, invisible in their free time) is still

R. Ricucci, *Ethnicity, Identity and Faith in the Current Migratory Crisis*,
https://doi.org/10.1007/978-3-030-84056-3_3

alive, overcoming other rhetoric (Bonikowwisky, 2016; Hall, 2017). This game seems to work until one crosses the Rubicon, which questions rights and customs, makes claims for recognition and safeguarding one's rights—an invisible (but real) barrier between what is perceived as being distant from, even irreconcilable with, the host society. The limit has been breached in recent years by asylum seekers portraying themselves as a challenge to the declarations of welcoming and hospitality, which innervate many countries' constitutions (Crawley & Skleparis, 2018).

From many points of view Muslims too exceed the limits within which people are commonly prepared to accept (or at least tolerate) the presence of the immigrant "other" (Massoumi et al., 2017).

3.1 Rediscovering Cities as Multireligious Arenas

> "Pope Francis from his first gesture, in Lampedusa, invited everyone to welcome him. No one excluded. Moreover, every time he spoke about migrants he recalled the respect for the religion of all, inviting on several occasions to collaborate with Muslim organizations. Yet, even in Catholic circles there is resistance to becoming aware of living in a multi-religious society" (Catholic priest, Barcelona).

> Walking in the street you can perceive how migrants have transformed the urban landscape of the neighbourhoods. Grocery shops, clothing shops, travel agencies, bureaucratic agencies are some examples of a varied list of activities that, especially in some neighbourhoods, have reshaped the range of services available. Added to this are the sounds that multilingualism produces, the smells of ethnic food shops and restaurants. All this is well known and translated into multi-ethnic city guides. But when you start talking in schools, in public debates about the presence of Catholic churches that have now become churches of the Romanian community or the musalla issue, the mood changes. There is a growing discontent against those who want to transform 'our religious landscape'. The audiences (often made up of atheists or non-practising believers) find themselves united in defending an alleged attack on religious homogeneity (M, educator in an intercultural centre, Turin).

The migrants have brought religion back to us. Through them, religion has become visible again: the processions in the streets or the 'mutton' festival organized in open-air places, the ceremonies of the Orthodox calendar bringing together thousands of people in and out of churches or the festival of lights. Events that underline how little support there is for the secularization thesis that religion will disappear within late modernity. In addition, the migrants' perspectives fit well into Hirschman's statement that "it is only through religion, or spiritual belief, that many people are able to find solace for the inevitable human experience of death, suffering and loss" (Hirschman, 2004, p. 1207). Indeed, in relations with host societies, religion can play a significant role in dealing with the implications of leaving one's home country.

Religious complexity is increasing in Europe and—when it overlaps with migratory flows—it is one of the key factors which affects media debates on multiculturalism. Indeed, the new centrality of religious discourse in the public sphere deals with the visibility of religious ethnic communities and their requests of visibility and recognition: the academic debate focuses on policies recognizing and safeguarding rights as well as those of social inclusion and welfare, which characterized the second half of the twentieth century (Giordan & Pace, 2014).

The presence of communities of Muslims (and also Sikhs and Pentecostals) in addition to Orthodox Christians has become more visible as a consequence of migratory flows to Western Europe, especially after the Second World War. However, plurality of faiths is a historical theme on the European scene. The renewed social blend resulting from mobility processes has once again placed centre-stage a theme which for many people belonged in history books or fuelled philosophical debate. Thus, migration—through its actors and at all times—forces societies and institutions to deal with variegated actors and untried ways of managing social phenomena (Table 3.1).

These transformations are often a source of anxiety and anguish for autochthonous populations, sometimes leading to alarmist reactions (Pickel, 2019). In the face of these reactions, local authorities try to find solutions—hitherto extremely heterogeneous—even regarding religious affairs (Becci et al., 2016).

The fact is that today there is no European model for managing religious diversity; the countries that for some time have had a high proportion of resident foreigners have undergone quite diversified experiences (Dikici, 2019).

Although relations with Islam are not a novelty in Europe, the discovery of the nexus between religion and immigration, and its socio-economic as well as political-cultural repercussions, has achieved growing importance as a result of historical events. First among these was the 1983 march of the *Beurs* in France. For the first time second-generation Algerian and Maghrebi immigrants took to the streets to make their voices heard and protest against their being treated as if they had just arrived. Whereas their parents had been capable of putting up with harassment and discrimination, they—born and grown up in the homeland of *égalité*—were not prepared to bend the knee to being considered replicants of the first generation, which is to say a reserve labour pool suited only to the most menial work and subject to assumptions about their origins and religion. To the extent that they were part of French youth, they demanded the right not to be discriminated

Table 3.1 Phases of relations with religious identity across generations

Belong and believe	Belong and... sometimes believe
1G	2G
In the shadow, without requests	'Do it yourself' and challenge the religiousness of parents
Tension (and attention) to be 'good guests'	Religion as cultural landmarks
Between 'pride' (asking for dignified places of worship) and 'prejudice' (being under observation)	Moving away from the lived religiosity which characterizes migrants
Involving young people 'under certain conditions' (useful as organizers, far away from the decision-making process)	Increase in the number of nones

Source: Table based on Ricucci, 2014, 2019

against. Although three decades have passed since then, in many scenarios little has changed (Gonzalez & D'Amato, 2017).

Another event, which at first sight seems to have no connection with Europe-Islam or immigration was the 1988 fatwa decree by Ayatollah Khomeini against Salman Rushdie, the author of Indian origin who wrote the Satanic Verses. Rushdie lived near Manchester, England, where thousands of Muslims gathered to burn copies of his book. Few today remember this happening, but it was important at the time because Europe discovered religious dissent within its borders. The leaders of the dissent were immigrants who had arrived decades earlier looking for work and had for some time been integrated into the socio-economic fabric of large or small English towns.

The following year, 1989, in Creil, France, another significant event took place. Three girls appeared at school wearing headscarves. On refusing to uncover their heads, they were expelled. This was the start of a national battle about the compatibility of religious symbols with the secularity of public spaces. A 2004 law put an end to the dispute by banning all religious symbols from state schools. However, contestation simmered on, the long saga concluding only in 2016 by a ban on wearing burqas on the beaches of the Côte d'Azur. All the above-mentioned events seem to belong to another era. Indeed, for some people (especially for the Millennials), discussion of Khomeini, the French episode on the veil could be obscure if we compare it with the latest terroristic attacks, which have strongly fitted in the issue of Muslims in Europe.

Attacks that occurred in Madrid, London, Nice, Copenhagen, and Brussels, are examples: "All the episodes have focused intense attention on the results of the integration processes, as well as on the relations among different cultural demands, making dangerous spectres of intolerance and discrimination re-emerge against those who come from Muslim countries (Cesari & McLoughlin, 2005; Crul, 2016). Therefore, in a climate of suspicion, the children of immigration, sometimes European citizens, have come under observation again, revealing that the granting of citizenship is not enough to avoid processes of differentiation and discrimination related to cultural origin. On the contrary, these conflicts, which have exploded in the urban peripheries, are commented on as a further indicator of the need for urgent reflection on the effectiveness of integration policies and on the destinies of the second generations (Emerson, 2009; Triandafyllidou, 2016)" (Ricucci, 2021, p. 12).

Beyond scientific analysis, over a quarter of a century, several incidents—some tragic—have demonstrated not only the inextricable link between the sacred and immigration but also the vitality of the religious (at least in the Muslim community) factor in a geographical context where the secularisation mantra is repeated with intensity only to yield later to a reality where *God is alive and well* (Copson, 2017; Becci et al., 2013; Beckford, 2012).

If native-migrant interaction occurs at least on the linguistic level, an ethnic and national element characterizing all of us, what happens with religion?

3.2 Migrants, Ethnicity, Faith

José Casanova (1994, 2018) pointed out the dissemination of and, at the same time, the refusal to limit religion to private space. The Spanish-American sociologist was observing a phenomenon, which—in countertendency to the secularisation thesis—would once again place religion, with its demands for laicity from institutions,[1] and its leading actors, in the spotlight.

Some of these actors represent groups of believers linked to migrants who intervene in the public sphere, not only introducing elements of the religious (such as veils and processions, turbans and rosaries recited in the streets) but also interpreting their role as leaders to comprise safeguarding rights, supporting employment integration and becoming trusted interlocutors for local administrations (Bruce, 2017).

The relationship between immigration and religion, emerging from the shade to which it had been relegated, is becoming one of the links which must be considered because of its substantial impact on welcoming societies. The rapport is transformed, and may have significantly different consequences, to the extent that it depends on the characteristics of the various arrival flows. This perspective pertained mainly to the US where various scholars—before their European colleagues—investigated its various aspects and repercussions on relations with the autochthonous population. Indeed, over time the arrival of men and women from non-European nations has enriched the American debate further, shifting attention to religious needs beyond the Judeo-Christian tradition (Waldinger, 2015).

Researchers have sought to understand whether strong religious traditions (capable of orienting behaviour) improve the immigrant-integration process, especially among children, in welcoming countries. Alternatively, to what extent religious ties could help to be resilient with a positive impact on well-being; or investigating the role of faith in developing a transnational religious capital and to organize FBOs with peculiar advocacy goals from migrants' rights (Triandafyllidou & Magazzini, 2020).

The dispute then moved from the United States to Europe,[2] where the object of research became the comportment of Muslims, Sikhs and Hindus, particularly in England (Zanfrini, 2020). The relationship between foreign citizens and the sacred was approached through the first generations, rediscovering only recently the importance of young people's religious experience (partly as a result of dramatic terrorist acts, as hinted in the introduction), of their socialisation results on themes of the sacred and their decisions about how to balance the way of living of both the family and community, and nurturing faith and their personal itinerary—which has

[1] Consider, for example, church-state relations, agreements between administrations and various denominations concerning the management of religion in prisons, hospitals, the army and schools.

[2] Further themes inherent in migrations (from the results of second-generation integration paths to the role of ethnic capital, from analysis of ethnic enclaves to an in-depth study of political transnationalism) have been activated by trans-Atlantic reflection (Crul et al., 2012).

been shaped by culturally pluralistic environments (Triandafyllidou, 2018). These aspects characterize not only those who come from Muslim countries. Analysis of migratory flows towards Europe shows that routes have gradually shifted from North Africa first to Latin America and more recently to Eastern Europe, with migrations from Poland, Romania, Ukraine and Bulgaria becoming increasingly noticeable since the end of the 1990s. This process is even more evident in—relatively—recent immigration countries such as Spain and above all Italy (OECD, 2019).

Yet, despite the evidence of growing religious pluralism, European research concentrated on the increasing Islamic presence from different perspectives and through various lenses: religious faiths and practices; predictions of a certain kind of society, secular or Islamic; identity definition, Religious, National, Cosmopolitan; attitudes toward children's education and mixed marriages; and demands made of various European societies, recognition of feast-days, and teaching religion in schools (Bonino & Ricucci, 2021).

The way in which the cultural identities of foreign citizens change over time under the influence of the social context they are immersed in, and the nature of relations among themselves, are considered crucial themes in the study of the second-generation integration process.

3.3 Religious Places in Emigration: Bastions of Defence, Occasions of Memory

> You couldn't plan anything for Sundays. As far as our parents were concerned, going to Mass was more than a sacred duty—it was a mission, a question of honour. There was no way you could get out of being seen by relatives, neighbours from home and compatriots. And we children had to go with them. It was like being in a stadium: parents on one side and we children on the other; they were fans of everything remaining unchanged in time, showing that we could be the same in Milan as in Bucovina. We children, bored silly, wanted to escape from that atmosphere of incense, of sounds, prayers and gestures which do not belong to us, which take place only once a week, contradicting what we see on television, what we observe in the street, what we experience with our peers when we go to the parish recreation centre to play musical instruments or football (F, Romanian, 18, Rome).

> My mother told me that when she came here she spent every moment of her spare time in the Philippine chaplaincy. She felt at home, reassured. At a certain point a priest arrived who had lived in the Philippines, so he spoke the language well. She was happy about this, partly because she felt freer when going to Confession. Then, she recalls, they often organized dinners, celebrations, when they felt extremely sad. Life in Italy was not as easy as she had imagined, and work—consider that she worked as a housekeeper—gave her lots of problems, which she discussed with other Filipinas after Mass. In the parish, my mother found a group, a community, which she had lost when she left home. Then we children and our father came, but there was no skipping Mass on Sunday, or lunch in the parish premises or participating in their various other events. My sister sings in the choir. I get told off because I don't do anything. I want to be Italian because I live in Italy. As far as religion is concerned, I want to decide for myself whether, where and in what language I'll pray (M, Filipino, 20, Milan).

Whereas American analysis of the relationship between religion and migration (Hirschman, 2004; Hanley et al., 2008) has a long tradition, it is only quite recently (Foner & Alba, 2008) that scholars and policy-makers have begun to shine a light on the subject in Europe. However, this attention has been selective, focussing on one religion (Islam), some peculiar symbols (the mosque, the veil and, recently, the burkini) and on a few (religious or informational) functions carried out (Césari, 2012; Voas & Fleishmann, 2012). Yet, many aspects need to be highlighted in connection with the role of places of worship which have risen up as a result of migratory processes: from being a safe haven from insertion tempests to being an environment of refreshment for souls alienated when dealing with the trials and tribulations of rebuilding their life paths in a new context where diversity (skin colour, physiognomy, accent) obstructs access to work, neighbour relations and daily exchanges (Eurobarometer, 2019). Religious environments play a key role because, as Hirschman stated, "Immigrants, as with native born, have spiritual needs, which are most meaningful when packaged in a familiar linguistic and cultural context. In particular, immigrants are drawn to the fellowship of ethnic churches and temples, where primary relationships among congregants are reinforced with traditional foods and traditions. Immigrants also have many economic and social needs, and American churches, temples and synagogues have a long tradition of community service, particularly directed at those most in need of assistance" (Hirschman, 2004, pp. 1207–1208).

In addition to prayer and religious comfort, useful information can be found in places of worship to orientate oneself in the host society; there are services for job search and support for the neediest. However, there are also language courses, informal training courses to work as housekeepers or carers, for example. Alternatively, meetings are organized on the rights and legal practices related to migrants' lives.

These few examples are sufficient to demonstrate that places of worship organized by immigrants themselves, or supplied to them by private and public institutions, play a role in host countries, which is not only that of guarding the faith, of places to pray and keep up – outside one's homeland – the possibility of attending ceremonies and meeting religious people who share one's language and/or culture (Table 3.2).

Reflection on the role of ethnic churches (as well as mosques, musallas, Sikh temples and Evangelical and Pentecostal communities) takes us once again to the other side of the Atlantic (Gans, 1994; Portes & De Wind, 2004; Yep et al., 1998) where, as for studies of second generations, research on this subject began. As for

Table 3.2 Typology of relations between associationism and local institutions

		Level of involvement of immigrant associationism in city life	
		–	+
City attitude/type of activities promoted	Reactive	Informative	Consultative
	Interactive	Collaborative	Decision-making

Source: Re-adaption from Saksela-Bergholm (2009)

many areas of society, even for religion the perspective was assimilationist. Thus, also with regard to relations between the faithful and places of worship, the trajectory was from "a specifically dedicated environment, connoted linguistically and culturally", through various intermediate stages, to the progressive insertion of immigrants (and even more their descendants) into already-existing parishes, mosques, temples and synagogues. In line with socio-economic assimilation, the religious too was described through a so-called life cycle defined on the basis of the evolution of ethnic churches (Pasura & Bivand Erdal, 2016) but applicable to any place of worship.

The debate about "immigrant" environments of faith became polarized between two interpretations.

The first perspective was the conservative role in keeping the new national group coherent by means of recourse to the language and the customary way of organizing rites. Durkheim's spirit is hovering in the air, recognized even by those who are not very accustomed to sociology or who have consigned it to a corner of their memory: religious beliefs and ceremonies unite men and guarantee the social context necessary for the transmission of values and traditions. Various authors, while examining immigrant integration processes, have stressed that the role of the church is the most incisive of the various agencies promoting transnational links and the conservation of the culture of origin (Ebaugh & Chafetz, 2002; Wimmer, 2013).

A second perspective considers places of worship as a stage in an assimilation process, even from the point of view of the role and objectives of religious organizations. That is to say, as for behaviour, values and traditions, also for religious matters the passage from the first to successive generations is accompanied by progressive withdrawal from the necessity of ad hoc services (Philippine and Romanian parishes, Sikh temples and exclusively-Egyptian mosques) in order to achieve the insertion of immigrants into structures already present in the host society – structures that, in this sense, should themselves assume intercultural aspects. The choice that ethnic places of worship seem to face is between becoming intercultural and ceasing to exist.

Similarly, what happens in the transition from being a temporary resident to being a permanent new citizen. The reality is much more complex and heterogeneous than an ideal diagram can express. Weber had already taught that ideal types are useful conceptual tools for elaborating hypotheses and comparing concrete causes. The itinerary is not linear because every step presents problems. Among the new arrivals (the minorities), there could be anxiety and fear of being unwelcomed. On the other hand, regarding those who belong to the majority, there could be discriminatory attitudes towards new believers with a different way of believing and behaving. Indeed, believing in the same god could be expressed in various ways. The liveliness of religious expression that characterizes African churches, for example, seems to clash with the silence and private religion of churches in Europe. This opens the debate on the meaning of "opening the doors of churches – or mosques – to non-national believers." Even in places of worship, attitudes of 'ethnic defence' emerge.

Thus, places of worship are like mirrors of what is happening outside them in society, where many people still advocate the necessity of having separate services (not only for prayer but also for education, welfare and employment), a specific procedure for foreigners. Yes, foreigners, but many of the subjects of reference are no longer foreign or are no longer interested in maintaining such close bonds (including religious) with their original culture.

In fact, a growing number of young people of foreign origin are by now national citizens or going through the process. The question then, for society as well as for religious groups, is one already posed by some young people of foreign origin and children of parents who came from abroad:

"When can we stop carrying on our shoulders the burden of being children of immigrants? When will we be accepted? The passport is not enough because – even in churches – we are still children of immigrants. Was it like that in the past too with your migration?" The questioner is Roberto, a 23-year-old Peruvian boy studying at the Turin Polytechnic (a most renowned university in Italy), who has been offered a traineeship in a big computer-science company. Unfortunately, the answer is yes, Italian immigrants also went through the same experience. There is no need to go back to Italian-Americans or Italian-Australians: the same thing happened throughout Italy with the experience of internal migration.

Yet, when it comes to immigration, history and memory are not disciplines that are called into question. This is further confirmation of how the confrontation with the 'different', with otherness, is still something capable of provoking more *"gut"* than *"mind"* reactions. When religious diversity comes into play, migration of religious discourse and symbolic capital occurs in spaces not formally constituted as 'religious'.

3.4 The Daily Life of the Faithful: Creed and Associationism

Present migratory flows are rapidly modifying the features of neighbourhoods, the school-going population and the structure of small commerce.

Compared with a couple of decades ago, immigrants have introduced, willy-nilly, many profound changes into their countries of arrival, illuminating the frailties and short-circuiting of the best-known, most consolidated integration models. However, every transformation at local level must be viewed through a local lens – and it is in their reciprocal influence that changes in their relations with the religious, which can be grasped (Hanley et al., 2008). Therefore, it is necessary to embrace a broader perspective going beyond not only the "here and now" but also beyond "national and continental borders" and in this way avoid the risk of interpreting ideas, practices, behaviour and religious choices without inserting them into a larger framework. Again, other migratory experiences come to our aid: already at the time of early-1900 Polish migration to the US, described by the sociologists Thomas and Zaniek, the study of religion among the immigrants was carried out by interweaving a triple perspective (Vãsquez & Knott, 2014).

The first was that of everyday life in the context of emigration, how it was composed in a scenario where language, cultural references and practices were not always well received—sometimes misunderstood, often banned from public spaces such as school and work.

Then there was the backward glance towards their country of origin and the characteristics of religious socialisation where the first generations grew up, which they are not transmitting to their children without perhaps being sufficiently aware of the changed environment.

> My mother and my aunts keep complaining that the prayer rooms here [in Barcelona] are not like those in Casablanca. They continue to remember what they did when they were young, how the activities were organized, what they did, the roles that older and younger women had, the respect that there was for mothers and grandmothers. They do not want to understand that living in a country where Muslims are one group among many (and let's face it, not even the one most seen) makes a difference. Nor do they want to believe that things have changed in Morocco, too. It is up to us children to open the eyes of the first generations, even if we clash and decide to create our own associations. It is the only way to build a European Islam (M, Moroccan, 27, Barcelona).

Nor do they consider, as a key informant working in an intercultural centre in Montpellier said, "the force of living in a society where all sound at the same bit", i.e. using prayer formulae, sacred places and clothes with which the majority of the population identify and feel secure.

Finally, relations developed between the homeland and the countries of arrival were studied. It seems a paradox: to what extent religious identity and religious behaviour could affect the home country while daily life is spent so far away. For example, migrants of Catholic tradition arrive in Europe, above all in southern Europe, with many expectations, as recalled by Maria Lourdes, 49-years-old, from Cape Verde who emigrated to Lisbon and then moved to Rome:

> I remember the disappointment of discovering that religion was something forgotten in Europe. I thought everything was about religion and then even more so in Rome, in the city of the Pope. I felt lonely, often inadequate because I asked to be able to go to Mass on Sundays or to have the chance to attend Easter services. When I found out there was a parish where people from Cape Verde gathered, I felt relieved, and that I was back home. Many of us Catholics felt lost, even a little betrayed, by those very countries where we thought religion mattered most and was most visible.

The above three perspectives need to be inserted into the broader integration scenario: as has been hinted, religion is an important element in the insertion process and may affect its outcome (Garces-Foley, 2008; Lippert & Rehaag, 2012).

So what happens when one grows up in a setting which is different from the points of reference of one's parents concerning educational practice, the transmission of values and behavioural (including religious) demands? Moreover, what are the possible repercussions of this geographical-temporal lack of sync[3] on young

[3] Parents refer to a daily routine which has changed over time even in their countries of origin. The feeling of alienation of the migrant who returns to one's home country, described by Sayad (2002),

generations between the home environment and that outside, whether school or social activities? Can religious diversity obstruct integration?

According to the followers of assimilationist theory, integration presumes conformism even from the religious point of view—a thesis that is not shared by the scientific community (King, 1993). Recent research seems to testify to unprecedented vitality and vigour in religious associationism. If this is true, as Handlin (1973)[4] observed, that for migrants (at least the first generation) religion represents a safe haven and a psychological advantage for overcoming the "trauma of immigration" (Hirschman, 2014), today further elements need to be added when considering relations between emigration and the sacred.

By now, it is clear that religion is not destined to disappear from post-modern societies (Hjelm, 2015). On the contrary, as is demonstrated precisely by its importance for immigrant communities, religion continues to be an everyday part of life contexts on both individual and social levels. In addition, labour-market and welfare-system problems have accentuated the role of social resource and economic assistance played by many organizations of a religious nature (Mooney, 2013; Wimark et al., 2019).

Immigrants, like natives, carry with them not only economic necessities and social improvements for the second generations but also spiritual needs which can best be satisfied when one finds oneself in a comprehensible linguistic-cultural environment, recognized and familiar (Levitt, 2007). Given what has been said hitherto, it should come as no surprise that in various diasporas one comes across the establishment of ethnic churches, mosques and temples. They are places that meet the need for orientation and act as a "buffer" for the insertion of new arrivals into the welcoming society. Thus, they are spaces where various aspects of assistance—from the material to the spiritual—are combined, where first generations can freely express their religious and cultural values without being judged, discriminated against or stigmatized.

In Italy too, religious pluralism has been one of the main elements of social transformation in recent decades (Pace, 2013). As has been mentioned already, although the relevant debate has long dwelt on Islam, it must however consider how the landscape of the country has changed as a result of substantial migratory flows from Eastern Europe. Romanians, Ukrainians and Poles have profoundly modified the composition of the migrant population from the multi-religiosity point of view. Attention has however been concentrated on Muslims, who are not a majority of immigrants. Statistical data show that despite media attempts to inseminate fear of the risk of the "Islamization" of Italian society, it is not correct to speak of an "Islamic invasion".

The Italian situation may also appear exceptional to many observers because of the problems immigrants encounter in finding suitable places to practise their faith.

---

concerns the contrast with a reality that is different from that experienced before emigration and distance has deprived of its less edifying elements.

[4] Other authors – for example Lensky 1963, Smith, 1978 – have made interesting contributions in their studies of the 1960s US migrations.

Once again, a diachronic glance invites us to examine other migratory experiences with a view to what policymakers call policy learning: adopting an attitude of learning and analysis of how religious diversity is managed to evaluate whether experiences developed in other places can be replicated, thereby saving energy and money. Putting this operation into practice, it could have not only historical-documentary value but also deep and interesting applicative repercussions; one gathers relevant suggestions and ways of handling relations with the sacred, which are repeated outside one's homeland. Religious practice needs an "infrastructure" which translates into specially designated spaces, specialized "personnel", organization of ceremonies, and conciliation of religious and civil calendars. As in other areas of migrant insertion, in the religious field too it is a step-by-step process, starting from initial invisibility, leading to the presence of specific places of worship for single collectivities to attain proper recognition. National, ethnic and—later—intercultural associations play an important role in this itinerary.

Therefore, the religious experience of a migrant population is closely connected with the presence of multiple subjects and meeting places of an ethnic background (Berking et al., 2018).

Despite the importance of ethnic associations as a point of reference for immigrants, they often meet serious difficulties in remaining active; this is often linked to their dependence on volunteers and not being able to afford the necessary resources. Further critical factors should be added, especially with organizations often managed by people with no in-depth knowledge of what it means to run an association in Italy, which public institutions to turn to or how to acquire essential information and funding. In some cases, however, support (including financial) comes from the country of origin, linked to various tasks such as promoting the national culture abroad and assisting the insertion of migrants.

Cultural associations—with which faith-related activities are associated—also belong in this context. It should be stressed that a lively associative (both cultural and religious[5]) fabric is a key element in supporting the integration of migrants into Italian society (Sinatti, 2019). Here it is interesting to note that in recent years some of these organizations have begun to diversify their activities. Through social initiatives they function as lobbies for their communities in the various scenarios—mostly urban—where they are present.

Relations among different ethnic collectivities are generally weak: each group is seeking its own specific relationship with local institutions and society in general. In recent years, the necessity of improving the quality of collaboration and networking among ethnic associations at local level has been highlighted in various places. It has to be said that, because of ever-decreasing funds made available by public bodies, more and more initiatives are supported by different subjects working together.

[5] It makes sense to include all ethnic churches among religious associations. They may be informal aggregations, usually with Evangelical roots. It is interesting to note, for example, the growing presence of Pentecostal churches, an important point of reference for Africans, especially Nigerians (Scrinzi, 2016).

Unfortunately, these networks may turn out to be ineffective, particularly with regard to collaboration on integration policies. For many of these subjects, the reasons are not hard to find: the weakness of their representation with respect to the community of reference; the lack of representatives capable of speaking Italian and interacting on a theme, moving on from primary needs to other—often more complex—topics; and the presence of many Italians involved in intercultural associations, often most positive in terms of involvement but often greatly obstructing the emergence of new leaders among migrants themselves.

In this situation, as we have seen, religious associations are often the principal point of reference for many immigrants, particularly at the beginning when language problems, the necessity to obtain useful information and a degree of suspicion of public services, different from those in their home countries, encourage attendance at places where they can find compatriots – parishes, mosques, prayer halls and informal meeting-places (Zamponi, 2017; Ambrosini, 2020).

3.5 Unwelcomed Believers in the Pope's Country

Despite the undoubted centrality of Catholicism, Italy is a culturally and religiously plural country. After years of debate and doubts about this definition, today it is no longer an opinion but is solidly supported by reliable statistics (IDOS-Confronti, 2020).

In 2020, Italian towns and cities may be described from the perspective of diversity of origins, languages, cultures and relations with the sacred, which do not identify with either Catholicism or the historical presence of Jews and Waldensians. This is not a recent phenomenon but already towards the end of the first decade of the twentieth century traces of marked differentiation were detectable: over three decades of migrations had profoundly redefined cities' social and cultural fabric; or at least this is the impression one gains from observing neighbourhoods whose commercial life is animated by ethnic shops and the high proportion of new enterprises set up by immigrants. On the other hand, as we shall see, full awareness of this situation has not yet sunk in. The collective imagination interprets the phenomenon mainly through emotions and reactions of closure towards the most communicatively striking events and an often-difficult, multifaceted cohabitation.

Therefore, the national panorama is becoming increasingly complex from the point of view of relations with the sacred. There is no doubt that immigration has contributed to replacing the religious-pluralism topic: this element existed before the flows that since the 1980s have turned Italy into a multicultural country. However, the magnitude of migratory flows has transformed this aspect into a central question in the life of Italian towns and cities—up until the 1990s it was mentioned only in ministerial documents or in those concerning the Catholic Church's inter-religious dialogue. Yet relations with religious diversity still present problems.

In Rome, in 2016 (and on subsequent occasions), an Islamic association organized a demonstration in front of the Colosseum, in Rome, another symbolic site not just for Italy but at the ancient heart of Europe. They were demanding the right to have their own places of worship.

In Italy—and specifically in the city of Turin[6]—in 2019, Islamic associations gathered in an open-air space, a former industrial setting, for the final *Eid el Fidr*. Civil and religious authorities joined the event: it was an example of inter-religious dialogue in practice as well as an image of public acceptance thanks to the presence of the city's mayor and other key administrative players (Mezzetti & Ricucci, 2020).

Alongside these images (symbolic and, for many, provocative), the Internet and means of communication were populated by Muslims praying in squares, buildings, suburban sites and open spaces granted by local institutions—but always far from the city centre.

We have already said that there is no doubt that in recent decades Islam has catalyzed the interest of scholars, journalists and public administrators. Less attention has been paid to religious affiliations, which have been part of the Italian scenario for centuries, or to those considered the offspring of a society inserted into global-influenced phenomena (Buddhist, Hindu) and multicultural dynamics (Sikh, Pentecostal). One therefore asks if there is any interest in knowing the outlines of the beliefs and rituals, which inform the cultural life and daily routine of many residents in Italy. In 2009, 29% of interviewees expressed a desire to learn more about religions. A decade later the proportion was more-or-less the same (33%) while – judging from relevant surveys—there was a marked increase in the possibilities of getting to know more in depth the various doctrines and experiences. In addition to, and apart from, numerous scientific debates, over the years various initiatives have presented the variegated Italian reality from the religious perspective: among these, visits to open days of mosques, synagogues and temples; school and neighbourhood projects; and the organization of festivals.

However, this increase in cultural offers did not scratch the surface of considerable Italian indifference to the subject. This outcome should occasion no surprise considering the growing climate of (national, ethnic and gender) identity revival: identifying with, and regrouping around, a strong identity element is indicated by many as today's key in dealing with otherness (Bayrakl & Hafez, 2018). For some it is national or regional rediscovery, even on cultural and political planes; for others it is the element of gender or sexual orientation; and for still others it is belonging to a religious group.

Among different countries and ethnic groups, the various generational groups have different ideas regarding their religion (beliefs and practices) and they live by them in their own ways. Age, sex, parental status (grandparents, parents, children), the socio-economic profile and migratory experiences matter. Of course, to what extent and which kind of societal environments they interact with play a strong role in defining religious identity. For all religious traditions, immigrant status and immigrant background mean reshaping both religious orthodoxy and rituals for

[6]Turin – among the Italian cities – represent an interesting setting for studying both integration policies and religious diversity management (Saint-Blancat & Schmidt di Friedberg, 2005; Saint-Blancat, 2019; Mezzetti & Ricucci, 2020).

several reasons: sometimes, due to lack of suitable places of worship; in other cases, to avoid conflict with the host society by presenting the image of communities that are not interested in gaining power in the public arena. Accommodation is not easy and sometimes it means internal conflict (Yilmaz, 2016).

> We often compare ourselves with younger people; young people, in a manner of speaking. We are between the ages of 28 and 35, we are said to have families and parents, but for our elders, the first generations, we are 'too young', 'inexperienced". What they mean is that, not having grown up in a Muslim context, we are "Muslim but not 100%". This is like adding water to wine to make it less strong, to cite an Italian example. Here are our 'elders' who think that we want to 'water down the religion' in order to respond to the requests of those who are afraid of Islam. The conflicts are numerous and the risk is that the younger ones, even if they feel religious, will move away from the prayer rooms. (M, representative of a Moroccan Association, Turin).

> Young people do not understand. They think we want to control them. We have known fatigue, contempt, racism because we prayed, because we fasted. For our children, whom we have spared these humiliations and for whom we have worked and sacrificed ourselves, everything seems easy. We are and will always be considered different. Many of us now have Italian citizenship, but we continue to feel watched on the buses if we wear the hijab or dishdasha or to suffer comments every time there is an attack. (M, representative of an Egyptian association, Turin)

In other words, the migratory experience imposes the need to restructure religious boundaries and codes of interaction with the host society on a different basis from the way things used to be done in their home countries in order to reinforce and maintain their religious identity. In different ways, this occurs in all the faith communities, from Muslims to Orthodox, Sikhs to Catholics, Pentecostalists to Buddhists.

References

Ambrosini, M. (2020). The Urban Governance of Asylum as a "Battleground": Policies of Exclusion and Efforts of Inclusion in Italian Towns. *Geographical Review*. https://doi.org/10.1080/00167428.2020.1735938

Bayrakl, E., & Hafez, F. (2018). *Islamophobia in Muslim majority societies*. Routledge.

Becci, I., Burchardt, M., & Casanova, J. (Eds.). (2013). *Topographies of faith: Religion in urban spaces*. Brill.

Becci, I., Burchardt, M., & Giorda, M. C. (2016). Religious super-diversity and spatial strategies in two European cities. *Current Sociology, 5*, 1–18.

Beckford, J. (2012). Public religions and the postsecular: Critical reflections. *Journal for the Scientific Study of Religion, 51*(1), 1–19.

Berking, H., Steets, S., & Schwenk, J. (Eds.). (2018). *Religious pluralism and the city: Inquiries into postsecular urbanism*. Bloomsbury Publishing.

Bonikowwisky, B. (2016). Nationalism in settled times. *Annual Review of Sociology, 42*(1), 427–449.

Bonino, S., & Ricucci, R. (2021). *Islam and security in the west*. Palgrave Macmillan.

Bruce, T. C. (2017). *Parish and place: Making room for diversity in the American Catholic Church*. Oxford University Press.

Casanova, J. (1994). *Public religions in the modern world*. University of Chicago Press.

Casanova, J. (2018). Globalizing Catholicism and the return to a "universal" church. In S. H. Rudolph (Ed.), *Transnational religion and fading states* (pp. 121–143). Routledge.
Césari, J. (2012). Islam: le défi à la sécularité européenne. *Socio-anthropologie [online]*, 25–26.
Cesari, J., & McLoughlin, S. (2005). *European Muslims and the secular state*. Ashgate.
Copson, A. (2017). *Secularism: Politics, religion and freedom*. Oxford University Press.
Crawley, H., & Skleparis, D. (2018). Refugees, migrants, neither, both: Categorical fetishism and the politics of bounding in Europe's "migration crisis". *Journal of Ethnic and Migration Studies, 44*(1), 48–64.
Crul, M. (2016). Super-diversity vs. Assimilation: How complex diversity in majority–minority cities challenges the assumptions of assimilation. *Journal of Ethnic and Migration Studies, 42*(1), 54–68.
Crul, M., Schneider, J., & Lelie, F. (2012). *Second generations in Europe*. Amsterdam University Press.
Dikici, E. (2019). Governance of religious diversity in Western Europe. *Ethnicities*. https://doi.org/10.1177/1468796819832342
Ebaugh, J., & Chafetz, H. R. (2002). *Religion across borders: Transnational immigrant networks*. AltaMira Press.
Emerson, M. (Ed.). (2009). *Ethno-religious conflict in Europe. Typologies of Radicalisation in Europe's Muslim Communities*. Centre for European Policy Studies (CEPS).
Eurobarometer. (2019). *Europeans' opinions about the European Union priorities*. European Commission.
Foner, N., & Alba, R. (2008). Immigrant religion in the US and Western Europe: Bridge or barrier to inclusion? *International Migration Review, 42*, 360–392.
Gans, H. J. (1994). Symbolic ethnicity and symbolic religiosity: Towards a comparison of ethnic and religious acculturation. *Ethnic and Racial Studies, 17*(4), 577–592.
Garces-Foley, K. (2008). Comparing catholic and evangelical integration efforts source. *Journal for the Scientific Study of Religion, 47*(1), 17–22.
Giordan, G., & Pace, G. (Eds.). (2014). *Religious pluralism. Framing religious diversity in contemporary world*. Springer.
Gonzalez, F., & D'Amato, G. (Eds.). (2017). *Multireligious society: Dealing with religious diversity in theory and practice*. Routledge.
Hall, S. M. (2017). Mooring "super-diversity" to a brutal migration milieu. *Ethnic and Racial Studies, 40*(9), 1562–1573.
Hanley, L. M., Ruble, B. A., & Garland, A. M. (Eds.). (2008). *Immigration and integration in urban communities. Renegotiating the city*. Woodrow Wilson International Center for Scholars.
Hirschman, C. (2004). The role of religion in the origins and adaptation of immigrant groups in the United States. *International Migration Review, 38*(3), 1206–1233.
Hirschman, C. (2014). Immigration to the United States: Recent trends and future prospects. *Malaysian Journal of Economic Studies: Journal of the Malaysian Economic Association and the Faculty of Economics and Administration, University of Malaya, 51*(1), 69–85.
Hjelm, T. (2015). *Is god back? Reconsidering the new visibility of religion*. Bloomsbury Academic.
IDOS-Confronti. (2020). *Dossier statistico immigrazione*. IDOS.
King, R. (Ed.). (1993). *Mass migration in Europe: The legacy and the future*. Belhaven.
Levitt, P. (2007). *God needs no passport: Immigrants and the changing American religious landscape*. The New Press.
Lippert, R., & Rehaag, S. (Eds.). (2012). *Sanctuary practices in international perspectives: Migration, citizenship and social movements*. Routledge.
Massoumi, N., Mills, T., & Miller, D. (Eds.). (2017). *What is islamophobia? Racism, social movements and the state*. Pluto Press.
Mezzetti, G., & Ricucci, R. (2020). Political opportunity structures and the activism of first- and second-generation Muslims in two Italian cities. In J. Martinez-Arino (Ed.), *Governing religious diversity in cities: Critical perspectives* (pp. 42–59). Routledge.
Mooney, M. A. (2013). Religion as a context of reception: The case of Haitian immigrants in Miami, Montreal and Paris. *International Migration, 51*(3), 99–112.
OECD. (2019). *Ready to Help? Improving resilience of integration systems for refugees and other vulnerable migrants*. OECD Publishing.

Pace, V. (2013). Achilles and the tortoise. A society monopolized by Catholicism faced with an unexpected religious pluralism. *Social Compass, 60*, 315–331.

Pasura, D., & Bivand Erdal, M. (Eds.). (2016). *Migration, transnationalism and catholicism*. Palgrave Macmillan.

Pickel, G. (2019). *Weltanschauliche Vielfalt und Demokratie. Wie sich religiöse Vielfalt auf die Demokratie auswirkt*. Bertelsmann Stiftung.

Portes, A., & De Wind, J. (2004). A cross-Atlantic dialogue: The progress of research and theory in the study of international migration. *International Migration Review, 38*(3), 828–851.

Ricucci, R. (2014). *Second generations on the move in Italy*. Lexington Books.

Ricucci, R. (2019). Religious belonging in family, school, and ethnic communities: Changes in Christian–Catholic second generations in Italy. In E. Arweck & H. Shipley (Eds.), *Young people and the diversity of (non) religious identities in international perspective* (pp. 33–50). Springer.

Ricucci, R. (2021). Managing the Muslim diaspora: Comparing practices and policies in several European cities. In R. Ricucci & S. Bonino (Eds.), *Islam and security in the west* (pp. 1–20). Palgrave.

Saint-Blancat, C. (2019). Introduction. *Social Compass, 66*(1), 3–23.

Saint-Blancat, C., & Schmidt di Friedberg, O. (2005). Why are mosques a problem? Local politics and fear of Islam in Northern Italy. *Journal of Ethnic and Migration Studies, 31*(6), 1083–1104.

Saksela-Bergholm, S. (2009). *Immigrant associations in the metropolitan area: Forms of mobilisation, participation and representation*. University of Helsinky.

Sayad, A. (2002). *La doppia assenza*. Raffaello Cortina Editore.

Scrinzi, F. (2016). *Latin American migration, evangelical Christianity and gender in Italy*. European University Institute.

Sinatti, G. (2019). Humanitarianism as politics: Civil support initiatives for migrants in Milan's hub. *Social Inclusion, 7*(2), 139–148.

Smith, L. (1978). Religion and ethnicity in America. *The American Historical Review, 83*, 1155–1185.

Triandafyllidou, A. (2016). *The Routledge handbook of immigration and refugee studies*. Routledge.

Triandafyllidou, A. (Ed.). (2018). *Multicultural governance in a mobile world*. Edinburgh University Press.

Triandafyllidou, A., & Magazzini, T. (Eds.). (2020). *Routledge handbook on the governance of religious diversity*. Routledge.

Vãsquez, M. A., & Knott, K. (2014). Three dimensions of religious place making in diaspora. *Global Networks, 14*(3), 326–347.

Voas, D., & Fleishmann, F. (2012). Islam moves west: Religious change in the first and second generations. *Annual Review of Sociology, 38*(1), 525–545.

Waldinger, R. (2015). *The cross-border connection: Immigrants, emigrants and their homelands*. Harvard University Press.

Wimark, T., Haandrikman, K., & Meinild Nielsen, M. (2019). Migrant labour market integration: The association between initial settlement and subsequent employment and income among migrants. *Geografiska Annaler, 101*(2), 118–137.

Wimmer, A. (2013). *Ethnic boundary making: Institutions, power, networks*. Oxford University Press.

Yep, J., Cha, P., Cho Van Riesen, S., Jao, G., & Tokunaga, P. (1998). *Following Jesus without dishonoring your parents: Asian American discipleship*. InterVarsity Press.

Yilmaz, F. (2016). *How the workers became Muslims: Immigration, culture, and hegemonic transformation in Europe*. University of Michigan Press.

Zamponi, L. (2017). Practices of solidarity: Direct social action, politicisation and refugee solidarity activism in Italy. *Mondi Migranti, 11*(3), 97–117.

Zanfrini, L. (Ed.). (2020). *Migrants and religion: Paths, issues and lenses*. Brill.

Chapter 4
Being a Muslim in the Diaspora: New Opportunities and New Challenges

Thinking of religion in emigration often means drawing a clear line dividing the religiosity of the first and the second generations, often more secularized. However, it is useful to recall here that results of recent empirical studies (Crul et al., 2012) have suggested that religion plays a significant role in the lives of young second generations, contrary to the experience of European immigrants of the past centuries (Kmec, 2018).

Often, the first generation, being in a new reality, without familiar linguistic-cultural references, favoured seeking refuge in religion, perceived as an element of recognition and reinforcement of identity, and as a cure for loneliness and isolation. This is mainly because attendance at mosques and places of worship allowed meeting with other migrants coming from the same country with whom they could share experiences and needs (Bastenier & Dassetto, 1993).

In fact, the mosques (or rather, the prayer halls) still continue to be a point of reference for the first generations, carrying out those functions typical of religious organizations in emigration, not only as religious reference, but also (and especially) as an identity reference (Bonjour et al., 2011). In this analysis, however, it is important to highlight that mosques or prayer halls, while constituting an important religious hub—and in some cases central to the cultural life of Muslim immigrants—cannot be considered as the only channel for the transmission of linguistic, cultural, and identity values. In fact, associations, whether formal or informal, national or transnational, have played and still play an important role.

Transnationalism refers to the ability of many immigrants to be active in the country of origin as well in the host country, and to maintain social, economic, political, cultural and religious relationships between the two contexts (Ambrosini, 2014). Scholars of migration have also observed how flows of migrants are very often accompanied by the emergence of new ethnic religious institutions that, as well as providing spaces for migrants to maintain the religious beliefs and practices they left "back home", also offer them financial, social and psychological support in their adaptation to a new environment. As well as exploring the positive contribution of these religious institutions in the lives of migrants, and the ways in which

R. Ricucci, *Ethnicity, Identity and Faith in the Current Migratory Crisis*,
https://doi.org/10.1007/978-3-030-84056-3_4

they facilitate transnational connections, some scholars (Cadge & Ecklund, 2006; Van Tubergen & Sindradóttir, 2011) have also shown their potentially negative role, in particular with regard to the pressure that some may place on migrants to conform to a certain set of practices, and even make financial contributions in exchange for the promise of salvation.

In religious practice, there may be important generational differences, for all nationalities. There are cases in which religious socialization can result in a younger generation that continues the tradition of behaviour and religious practices of their parents. Others, in the constant search for balance, end up assuming different roles and behaviour in the family and with their peers (Ricucci, 2017). In other cases, socialization gives rise to processes of differentiation and distancing: autonomous paths of relation with the sacred.

For the second generation, religion appears to satisfy the desire to find a sense of moral certainty and their place in the world (Roy, 2002, 2007), highlighting their originality within a precise membership/belonging. Islam, practised together with their peers, offers second generations stability and a framework in which to live as Muslims in European society (Conti, 2018) while continuing to question both with respect to the religious identity of their parents, and with respect to the perceived religious identity in their countries of origin, in which being a Muslim is not subject to stigma and difference and certain behaviour does not require continuous justification.

Also for the second generation, they sometimes suffered discrimination—the questions of their classmates and, in recent years, curiosity and attention to their countries of origin (e.g. during and after the Arab Spring in Egypt and Tunisia) and to their religion (due to terrorist attacks in Europe). These are all factors that have stimulated new reflections about their personal identity, and personal and collective religious affiliation. In fact, young people often choose "their way, looking for compromises and new syntheses, asking questions about the traditions they have inherited, living faith in a personal and authentic way and not uncritically" (Granata, 2010, p. 88).

Considering the impression children have of the religious context of the family, there is the start of a process of revision and reinvention of religious practices among second generations, which reflects a growing subjective autonomy in contrast to the inherited tradition. The different and complex ways of young people for dealing with Islam therefore have a common denominator: progressive detachment from ethnic Islam and the affirmation of individual logic in their relationship with the religious dimension. It is a process taking place in some Muslim societies, but in Europe it is much more evident. It is important to emphasize that this process of individualization of choice and religious practice is a factor of decisive cultural convergence within European culture in reference to religious belonging, which is facilitated by the Internet and social networks as spaces where a believer can find many different ways and examples of being Muslim, sometimes contradictory, and choose what they think is the most appropriate for them. The individualization of religious choice is further favoured in Europe by the absence of a "legitimate" and recognized Islamic religious authority.

Table 4.1 Muslims in the EU, Norway and Switzerland. Estimated size of Muslim population in 2016 and in 2050 with a zero-migration scenario. Total and first main countries according to the rate of resident foreign population

	2016	2050 under zero migration scenario
Region (total)	25,770,000	35,770,000
France	5,720,000	8,600,000
UK	4130,00	6,560,000
Germany	4,950,000	5,990,000
Italy	2,870,000	4,350,000
Spain	1,180,000	1,880,000

Source: Pew Research Center 2017.

If it is true that the integration of the Muslim population in Europe is a major challenge both for European societies and for Muslims, as we saw in the previous chapter, it is becoming clear that young Muslims will play a key role in this process (Table 4.1). To understand the future prospects of the evolution of Islam in Europe, it is therefore becoming important to analyse to what extent and with what results Islamic belonging plays a role in the experiences of young people. However, it is becoming equally important to analyse the ways that religious belonging relates to other forms of belonging (Levitt, 2004), and the role of the Internet in these processes. The relationship between the first and second generations is going through a phase of increasing complexity: the growing differences in style and lifestyle choices, and links with the country of origin, its culture and religion, constitute an increasingly obvious challenge for the host society.

4.1 Elements of the Debate

> When my Romanian friend said that women go to mass wearing white veils, nobody reacted. On the contrary, the teacher said that this was the custom in Italy, too. When I said about the veil, a friend of mine immediately said that the veil is not allowed in Italy, that all women should be forced to take it off. Some girls also intervened. I tried to explain what the veil means for us, the idea behind it, the choice, but they wouldn't even let me speak. In the end, I banged my fists on the table and told them about the Muslim countries where you don't have to wear the veil, like Tunisia, about the deputies in Morocco, Muslim women who are not enslaved as many people think. There is a lot of ignorance and the school doesn't do much. There is little talk of Islam; there is no room for discussion. Moreover, even the teachers don't seem to want to speak. So, well, the Romanian students who don't create problems or those from Catholic countries, like the Ukrainian who came to my sister's class or the Peruvian girls I had. Everything simple, no flashy symbols, no special requests or parties that are not celebrated in Italy. In the end you have to survive: either you choose to be stronger and then do like my cousin, she is not afraid, she wears the veil, she argues, even in the street, when someone looks askance at her; or, choose my way, you shut up, you say nothing and try to be invisible (F, Marocco, 21, Turin).

For Muslims, at the centre of attention are the symbols, traditions and also the changes in emigration of a religion that every attack, far or near, is put under the spotlight, reserving to its protagonists—already European citizens or even still foreigners—the treatment of *"unwelcome guests to be treated with suspicion"*. Perceptions that are fuelled by the divisions that characterise the universe of Muslim associationism and stereotypes fuelled by the media that deal with the subject in a superficial way. The outcome is a distance taking, as the following Table 4.2 shows.

Islam in Europe, "Ties the threads of a memory broken by history. This means that the Muslim presence will be an additional reason to reflect, not only on the identity of these European Muslims, but also on the historical roots of Europe's identity and the fractures that have divided the Mediterranean basin for centuries" (Pace, 2004, p. 20).

On the other hand, European Islam not only shakes up identities and challenges the definition of invisible borders, which cannot be overcome by the acquisition of a passport (see Chap. 5), but also proposes the long-standing debate on the role of religion in the public sphere, "asking new questions and showing the partiality of established structures" (Monti, 2017, p. 37). How do the memberships of non-European national communities intertwine with the duties of every Muslim belonging to the Ummah, i.e. the Islamic supranational community? What, if any, are the irreconcilable issues between Sharia and the legal, economic and social systems consolidated in Europe and supported by rules and regulations, including international ones? More than 30 years after the first protest about the veil worn at school in France, numerous episodes have challenged the constitutional order of the major immigration countries (and therefore the presence of Muslims), highlighting the fragility of the structures that regulate the relationship between Churches and public institutions. The religious calendar, school teaching, the construction of places of worship, the recognition of ministers of worship in public places, such as hospitals, barracks, prisons; the possibility of wearing specific clothing, prayer times, the legal recognition of organisations and their financing: these are all issues that call into question a construction with a precarious foundation. Here the Muslim plays a mirror function, showing the aporia of compromises and the anachronism of institutional architectures challenged by human mobility and the progressive stabilisation

Table 4.2 Percentage of respondents who would not like to have... (% response in the main EU countries of old and recent immigration)

Country	A Muslim in your family	A Muslim as a neighbour
Italy	43%	21%
United Kingdom	36%	16%
Austria	34%	13%
Germany	33%	13%
France	24%	11%
Portugal	20%	11%
Spain	18%	11%
Belgium	14%	7%
Sweden	14%	8%

Source: Pew Research Centre, 2017

and structural presence of immigrants who—by conviction or cultural adhesion—have a reference point in Islam for religious, economic and social life.

4.2 Attempts at Religious Cohabitation. Towns and Cities in the Mirror

There is a kind of step-by-step path that leads Muslims in Europe from being silent workers to becoming activists in demanding full respect for the freedom of religion inscribed in the constitutions of democratic countries. In the context of the various European towns and cities investigated, it is possible to retrace these stages, which are summarised in Table 4.3 below.

The change of register is significant: as Scholten (2011) argued, we go from being the recipients of interventions to becoming co-protagonists in policymaking. In this sense, the intercultural declination, in the meaning of an "inclusive intercultural policy", pushes relations between Islamic associationism and local institutions to get out of the (explosive and reductive) dichotomy of immigrants vs. citizens. What is more, for second generations it is no longer a question of presenting demands that "confine" Islam to a question that is specific to immigrants, whose gaze is nostalgically turned to the past, but of inserting the discourse in the most desiderable debate on religious pluralism, freeing it from the link with immigration. Energy is invested in building partnership relations, to conquer areas of credibility and recognition: that is to say, to become reliable interlocutors of institutions and schools. In fact, the new actors often have no experience of migration, sometimes they are Italian citizens: that is, young people from another religion who seek space and express the will to participate in the intercultural and inter-religious policies of their town or city.

Table 4.3 Requests from Muslim associations at local and national level

Period	Actors involved	Requests at the local level	Requests at the national level
1980s	Converted Europeans and first generations	Ritual slaughter of animals, differentiated diet, **prayer** rooms	Recognition
1990s	Mainly first generations, representatives of associationism and the imams	Better **prayer** rooms, teaching of Islam at school, cemeteries	Understanding and agreement
2000s	1.5 generations, women, Muslim associationism	Mosques and requests for involvement in intercultural activities	Understanding and agreement
Since 2010	First and second generations	Requests for participation and involvement	**Respect of the constitutional right to religious freedom**

Source: author's elaboration on CLIP, 2010; Bonino & Ricucci, 2021.

Recently, with the arrival on the political scene of young people, the relationship between Islamic communities and local/national administrations seems to show a change in register and to catch the signs of a break in the ways and contents linking first- and second-generation associationism. The demands are more general, linked to recognition as actors and a significant part of the socio-cultural context of the town or city. Concerns are linked to activities to raise awareness and update citizenship on the ongoing generational changes in the Muslim presence. The terrain of "symbolic religiosity" (Gans, 1994), in which second-generation religious identities are only vaguely linked to beliefs and rites, but rather held together and strengthened by common belonging to an association. Through symbolic religiosity, belonging to Islam can be translated into the recognition of a common Muslim identity, shared and practised within associative activities, but not necessarily linked to the observance of practices. Therefore, there is dissociation between the practising Muslim and the one who recognises in Islam a cultural and identity reference. As regards this distinction, new demands and relations take shape (more on the side of collaboration and sharing than on the side of rupture and opposition) with the local reality. The aim is not so much the recognition *tout court* of one's own specificities, as much as the right to diversity and the promotion of intercultural policies, in which religious difference is one of the elements of the town's/city's social fabric and not a factor of conflict.

The protagonists of this phase are above all the second generations, at which many administrations have decided to aim, as key figures, to promote paths of social cohesion. In fact, rationally thinking about the paths of integration of young Muslims means deepening reflection on the ability of all the actors on the stage of relations between natives and migrants and their children (from teachers to operators of associationism). Finally, they represent an important test for intergenerational dynamics, within both the different ethnic communities and the citizenry as a whole, where the different integration paths of children of immigration challenge perceptions of adults and elderly people.

4.3 The Latest Frontier: Becoming a Transnational Muslim by Internet

Within a multi-cultural, multi-religious, interconnected, digital society, the use of the Internet for religious purposes has potentially important implications for inter-ethnic and inter-religious relations, policymaking and education. Furthermore, over the past decade, the Internet has provided a new platform for religious groups as a new approach and a venue for proclamation, communication and interaction, with both members and non-members of their religion. Literature is emerging on the ways in which the Internet intersects with religious faith, with many scholars postulating that the Internet may have potential significance in altering the conception and the practices of religious faith (Saunders et al., 2016) as a "modernizing" and "decentralizing" element in fundamentalist religions, which could potentially undermine them (Barzilai-Nahon & Barzilai, 2005), and promoting secularization (Armfield & Holbert, 2003). In this framework, little attention has been paid to the use of social networks among second-generation Muslims, but young Muslims are digital natives, always online (Prinsen et al., 2015), always connected with User Generated Content and linker people who express their indentity online and re-elaborate the world and the contexts in which they live using posts, pictures, videos, and gifs.

Online forums and social networks like Facebook are also places where people actively discuss truth claims or ask questions and find immediate and updated answers to problems concerning religious obligations and ethical concerns. The social networks can also help in the coordination and mobilization of social actions and in increasing visibility. In fact, they offer simple and less expensive ways to organize members, arrange meetings, and disseminate information and opinions. Digital technologies make new resources accessible to the young Muslims, also enabling them to assert their individuality and experiment with transnational practices. According to Leonini and Rebughini (2010, p. 18), the second generations in fact "grow up in a society that is, for the most part, transnational and globalized, where the needs for real integration and assimilation into a stated cultural model, closed within a nation's boundaries, are fading away". The Internet seems to have given impetus to these trends, increasing the opportunities for young people to consume "goods, pictures and representations, which have ever looser bonds with the nation state", providing global and transnational cultural and religious references. It enables the "claim of differences to take place now on a supranational linguistic and religious scale, with reference to tastes, aesthetics and traditions, which overcome the bonds of a state". Moreover, new media can also be significant in redefining present and future projects of first and second generations. The second generation has, in fact, unusually complex and ambiguous views on the home, identity and "where they belong". Thanks to the Internet, first- and second-generation migrants' "homes", or their notions of "here" and "there", are becoming less "topological" and more transnational and affective. Furthermore, media—technologies as well as contents—paly a part in defining the formative experiences of a generation, not only because they are so deeply embedded in everyday life that they become a "natural" element of the social landscape, but also because historical events, as well as cultural values and their symbolic forms, are often mediated by them (Silverstone, 2005). This happened, for example, with the events of the Arab Spring in 2012 as well as with the proclamation of the so-called Islamic State in 2014. Both were supported and information about them was disseminated, using new media, particularly social networks and platforms like Facebook, Twitter and YouTube, and instant messaging services.

There are risks and opportunities in the massive diffusion of new technologies and the use of social networks, blogs and Islamic forums, including the development of bottom-up interpretations of Islam, of an Islam called "cut and paste", eclectic, from which people can take inspiration according to their religious preferences. The Internet can also link young people born in Italy with religious practices distant from their experiences outside the family (in secondary socialization) and encourage the spread of visions, which are radical and irreconcilable with the context in which they live, producing situations of isolation and estrangement and bringing them more and more to seek refuge in "protected" environments online and offline, with the risk of being recruited to fight as members of an imagined global community. The Internet can be a channel of propaganda for radical Islamist groups and Western fighters (Hegghammer, 2010).

Research by Dounia Bouzar (2015), director of the Centre for Prevention Against Islamic Sectarianism, indicated that 91% of recruitment takes place online, but in many cases the Internet acts as a radicalization-accelerant in the further stages. It is estimated that there are nearly 50,000 Twitter accounts of ISIS supporters, with an average of about a 1000 followers each. Recruiters appeal to youngsters living in European countries and looking for their way of being Muslim with messages of martyrdom and loyalty packed in rock video formats. This was a new genre labelled *pop-jihadism* (Koehler, 2015). Recruiters convince young people that they can be part of something bigger that will allow them to forget about their "small problems" at home. In the digital age, jihadists can now reach people from different social backgrounds and demographic cohorts. That is what David Thomson (2014), author of the book, "The French Jihadists", called the "democratization of the cyber jihad". It is also noteworthy that, through social media like Facebook and Twitter, it is now easier to keep in contact and it may happen that those who left for Syria stay in touch with friends at home and chat after chat, get them to leave for the so-called Islamic State. As stated by researcher and expert of radicalization, Montasser AlDe'emeh (quoted in Birnbaum 2015, p. 1):

> If you play football together every day in the park and two of your friends go to Syria, you stay in touch with them on Facebook. In addition, they will say, 'Belgium is boring. Here we have beautiful rivers and Kalashnikovs. Here in Syria we are somebody'. In Belgium, you are nobody.

Contacts of this type have a strong potential for recruitment, even involuntarily or indirectly.

4.4 Muslim and Immigrant. An Italian Nightmare

In the country of origin of Pope Francis, himself a second generation of migrants abroad, the presence of different ethnic groups is neither a novelty nor a transitory phenomenon. Foreigners—and especially their children—have gradually become a structural presence in Italian society. Although the educational field continues to be the one to which most energy is dedicated, both in terms of resources and research, the second-generation avant-garde is approaching adulthood, going through the delicate transition phase after compulsory education to face a society which is often unable to distinguish between the paths of these young people (all Italian) and those of their parents (the result of one or more migrations). The children of immigration, whether born in Italy or arriving at a later point their life paths, are therefore a central element of the migration phenomenon.

As for the adults, who are present in a heterogeneous way in individual local situations, there are important territorial differences reflecting the different rhythms of migration flows as well as the effects of migration chains: in areas characterized by old arrival routes, rebalanced from the point of view of gender and with high reunification rates, there are more minors born in Italy. Conversely, among more recent arrivals whose integration projects may not yet be fully oriented towards definitive stabilisation in Italy, minors born abroad and reunited in adolescence prevail. However, the differences are not only those or place of birth. Further variables complicate migratory trajectories and make each one unique. Some of them are related to individual history: family type, socio-economic conditions, future prospects, and relations with both the community present in Italy and that of the home country. Others are more general, depending on the attitude of the host society towards immigrant citizens and their various origins, the way they are considered and the images/stereotypes that inform such consideration.

In this scenario, the question of religious affiliation assumes particular importance, in the sense that it represents an element of further differentiation in relations with Italian society and the opportunities it offers (Ricucci, 2017). In the background, we find ourselves faced with an Italy that is somewhat fearful of the growing religious pluralism resulting from migration flows. In fact, in a context that is already the common home of many religious expressions, the concerns focus on Islam: it is the Muslim man and the Muslim woman who pay a high price for international tensions and judgements about the growing numbers of believers whose faith is routinely perceived as conflicting with shared values considered fundamental in Western societies. Indeed they seem at times to be undermining the hard-earned gains to protect and safeguard human rights established after the Second World War. These rights include that of professing one's own religion. Daily experience in different social spheres tells us of mysterious "*attacks*", whispers, sideways glances, subdued comments and openly hostile attitudes endured by those who are perceived as being religiously "*unintegrated*", culturally "*unappreciated*", and socially "*untolerated*".

In order to understand better the Italian context, it is important to recall that Muslims in Italy mean mainly immigrants and not Italian citizens. Considering the relatively recent migratory phenomenon and the nationality law mainly based on the principle of *jus sanguinis*, the debate about Islam and its perception in Italy deals with immigrants' perception.

The official self-perception of Italy as an immigration country and the recognition of pending needs for immigrant integration in political and public discourse are determined by factors other than political changes in the Italian government in recent years.

Several scholars have stressed how integration issues need to be investigated at local level, where immigrants and Italian natives cohabit, and these are not hot issues for various reasons: (a) a city's policy of social regeneration in some areas where various ethnic groups cohabit and both social and religious conflicts could emerge; (b) local initiatives carried out by both public and private institutions aimed at promoting intercultural dialogue; c) An advanced integration process that means an increasing number of citizenship acquisitions, family reunions and second generations. In this framework the economic crisis between 2008 and 2016, which strongly affected Italy, seemed to push tensions against Muslims into the background: Italians were much more worried about their job situation than 'wasting time' in contrasting immigrants and their religious belonging (Ricucci, 2014).[1] In the most recent years, the situation has changed again and both migration and Islam has returned to be hot issues, supported by the strategies of the far-right Lega party (the League, formerly the Northern League) which put these two topics at the core of its political agenda (Anselmi & Blokker, 2019).

All the main Muslim organizations[2] working in Italy share a common interest in presenting a "moderate Islam". All the activities dealing with professing religion and organizing social activities are developed in collaboration with local institutions, while keeping the neighbourhood and its perceptions in mind (Saint-Blancat, 1997). For example, the openness of new places of worship (i.e. musallas) has been accompanied by activities aimed at explaining and presenting the Muslim Centre to the Italian residents living in the area. It is a specific interest of Muslim organizations to demonstrate their desire to promote integration. Interviews with Muslims carried out in several Italian towns and cities confirmed that all they really wanted was the freedom to practise their religion. They wanted Islamic festivities to be recognized and to be able to eat food according to the Koranic laws in public eating-places. They were concerned about getting their religion taught in schools, and about securing places to pray. These elements were more important to them than obtaining legal recognition from the Italian government.[3] In short, the desire to keep

[1] According to recent surveys, migration (i.e. border controls, increasing numbers of immigrants from Muslim countries) was not considered a hot issue: the economic crisis and unemployment were the main issues (Pless et al., 2020).

[2] The main one is UCOII (Union of Islamic Communities in Italy), a federation of about 60 mosques throughout the country. The *Centro Culturale Islamico d'Italia* (Islamic Cultural Centre of Italy) is the second biggest Muslim organization in Italy, mainly supported by Islamic States. It is not an association, but a religious legal entity, recognized by a decree passed by the President of the Republic. This Centre is Rome-based, at the Great Mosque, and promotes the official Islam of the States. The other two organizations (AMI—Association of Italian Muslims, and Co.Re.Is—Muslim Religious Community) are composed predominantly of Italian citizens who have converted to Islam.

[3] In Italy, the recognition and rights of religious minorities are regulated by bilateral agreements between the state and representative bodies of religious groups. Muslims have not yet succeeded in reaching an agreement with the state, although they constitute the second-largest religious group in Italy. According to Aluffi et al. (2008, p. 138), *"laws currently in force can give satisfactory answers to some of the basic needs of Muslims"*, e.g. mosques, cemeteries, ritual slaughtering and supplies of halal food. Without the agreement, however, Muslim communities cannot enjoy the benefits that such agreements bring. These include the right to abstain from work on religious holidays, to appoint teachers to public schools to provide religious instruction, to allocate a quota of personal income tax to the Muslim community, and to observe other religious rites.

their own religious tradition did not contradict their willingness to integrate and respect Italian law.

It is clear that the question of 'the versus' (up or down) of the socio-mobility process deals with various factors, among them the institutions of the host society that immigrants (especially children) must pass through, and the local contexts within the receiving society where immigrants live are important and have been neglected for a long time. Towns, cities and suburbs all vary in the size of their immigrant populations and they differ in how they manage requests coming from Muslim associations. Therefore, one might expect that the presence of people with a migration background is more accepted in regions where there is a higher level of social integration. In other words, "Is the fear and distrust towards something different, in this case towards Muslims, being reduced with the increase of integration?" Research findings show a rather complex scenario, which can be summarised in this way: "You are well accepted but NIMBY [Not In My Back Yard]". In fact, the regions where there is a higher level of acceptance are the ones in which the Muslim presence is not very significant and where immigration is considered more as an issue to be handled temporarily and not as a structural element of the context. Recalling Ricucci (2014, p. 75), "Northerners are more closed and Southerners more open and tolerant. Although this tendency confirms common sense, it has paradoxical aspects. In the North of the country, anti-immigrant and anti-Islamic feelings are undoubtedly more widespread despite the fact that in these areas the highest rate of labour market insertion of the foreign population is recorded". The Italian path towards integration seems to approach the concept of "segmented assimilation" defined by Portes. Even in regional contexts where there are higher rates of integration, i.e. residential and job insertion, civic participation, good knowledge of the language and citizenship, there are still some aspects that continue to mark the difference between citizens of Italian origin and foreign citizens or those of foreign origin. Religion, Islam in particular, counts.

4.5 Who Has the Power? The Case of the City of the Holy Shroud

Turin is located in the North-West of Italy, in the Piedmont region. The municipality has 908,902 inhabitants distributed across ten districts with varying population densities. It is Italy's fourth largest city and its history is closely connected with internal migration flows and the economic development of the 1960s, due principally to the automotive sector, led by Fiat (ex FCA, now Stellantis). For over 40 years, Turin has been a "mono-industrial" economy: economic crises and the consequent social problems characterized the late 1970s and the 1980s. The 1990s represented a turning point for Turin: a difficult transition from the Fiat system to a diversified economic structure not only favoured the development of new activities, but also contributed to investing in the promotion of the city's cultural and historical background, including being the site where the Holy Shroud is stored in its Catholic Cathedral. Along the same lines, the 2006 Winter Olympic Games represented a strategic opportunity to invest in infrastructure and improve the diversification of the economy by promoting international cultural events and developing the tourist sector.

Since the 1970s, Turin has received a constant flux of immigrants. The first included students, mainly from the Middle East, Senegal and Nigeria, and political refugees and opposition members from South America (Chile and Argentina), Eritrea and Somalia. The second wave, also in the 1970s, was composed mainly of domestic workers: initially Somali and Eritrean women, then Filipino and Cape Verdean nationals. The third wave occurred in the period between the mid-1980s and the mid-1990s: Turin witnessed a surge in the number of Moroccans and other Africans, Asians (Chinese and Filipinos) and Latin Americans (Peruvians). The last wave concerned Eastern Europe: Albanians first, then Romanians and others.

Since the beginning of its story as an immigrant city, Turin has been recognized as one of the first Italian municipalities to develop initiatives and projects to manage the increasing flows of migrants. In the last 30 years, the municipality has shifted from "action on demand", generally multicultural oriented, to a more coherent and specific intercultural policy.

This evolution has undergone four phases: the first-welcomed phase (1980–1990); the creation of a network of specific services based on interaction between the municipality and the third sector (1990–1995), implementing an interesting "welfare mix" model; the initial development of intercultural activities (1995–2005); and the development of intercultural policies (2006–2011).

In the first two periods, the policies can be defined as "multicultural", i.e. aimed at helping and promoting the first insertion of the various migration flows with special attention to each ethnic background. The interventions supported in that period were defined by the city administration as an ad hoc approach—that means that municipal initiatives were explicitly directed towards immigrants in specific difficult conditions, reacting to emerging social problems by mobilising all the main local actors, public and private, interested in the issue (e.g. language courses, information offices, shelter centres, and accommodation activities addressed especially to minors and women).

Attention to the second generations emerged mainly in the last period of the policy evolution, as described below. In fact, recently a new era seems to have come about: the consolidation of the intercultural discourse has taken place in tandem, generating a large number of practices, projects and experiences mainly based on the notions of dialogue, mutual exchange and social interaction. The centrality of this policy was confirmed by the creation, for the first time, of a city-level Department for Integration (2007), dedicated to defining a coherent intercultural policy for promoting integration. Other municipal departments have to discuss and collaborate with this new Department on each project, initiative and decision concerning integration issues. The goal is to insert the discourse on integration into all city policies, reinforcing the shift from special initiatives and ad hoc projects for migrants to policies capable of considering the various facets of the city's residents.

This shift was part of the last two mayoral programmes (2006–2011; 2011–2015), where it was considered necessary to develop the intercultural dimension as an approach affecting all policy areas and to promote the involvement of immigrants in the city's life in various fields: social, cultural, and economic. Immigrants should be metamorphosed from being (or being perceived as) recipients to being pro-active participants promoting activities.

In this phase, attention to second generations and their civic involvement came onto the stage. These young people were considered as the drivers of the integration process on both sides: on the one hand, supporting immigrants to be better inserted in the city and, on the other, helping Italian citizens in understanding the multiple aspects of immigration in the city.

This new approach towards juvenile activism has been supported financially in the framework of two calls for projects that the Municipality developed in agreement with a bank foundation. The initiatives admitted concern, mainly, second generations who had grown up in Turin and to whom the city turns in order to develop its own intercultural policies. The activities carried out in order to train second generations to become active citizens may be divided into three groups according to their functions:

1. Initiatives directed towards useful or practical assistance: concrete help by offering services of orientation and counselling;
2. Gathering emotional support activities, which may at times be defined as assistance towards self-expression and the formation of one's own personality. In this sense, we should also recall improving young people's communicative and expressive capacities and the consequent development of relational skills in free time;
3. Activities directed towards information and educational support to offer moments of updating, reflection and education on subjects relevant to minors and young people, to cope with educational challenges posed by adolescents, as well as to provide both young people and adults with useful information about educational and training paths.

The result of all these initiatives is the active involvement of young people as organizers, animators and educators of other foreign minors who are following insertion and growth paths in Turin. Meantime, these activities strongly involve second generations in the city's life by considering them active citizens of Turin, even if their citizenship is not Italian.

4.6 It Is Time to Change. New Actors Coming on the Stage

The generational passage can be noted from the relationship, which young people develop with religious belonging even more than practices and dynamics in the public arena.

To synthesize and simplify, young-Muslim interviewees can be divided into two groups.

First there are those for whom religion is little more than education received in the family: we can call them secularised young people who take part in some rituals, maintaining a collective Muslim identity in the face of relative indifference on the level of faith.

> It is easy for boys and girls not to follow their parents' teachings. Some who have come here—especially those who have married Italians—have forgotten Islam. Perhaps both parents work and the children go out with their friends, so they don't even speak Arabic. Fortunately, there are now aerials so that you can watch television from all over the world, which weren't there before. This allows the children to know Arabic and religion better. In our days, it all depended on us, but that was not enough. Now it's easier, partly because there are more young Muslim people. We brought our children up to be like ourselves, but they have their own ways. They have more Italian than Egyptian friends. When they ask my permission to do something, I leave them to it because they know their religion so they know what they can and cannot do (M, Egyptian, 56, Turin).

The interviewee stressed an important element of the relationship between young people and religion: the force of numbers. An increasing visibility of Muslim families, of the number of schoolchildren declaring themselves to be Muslim, girls wearing the veil and associational activity linked to religion may therefore become fertile terrain for the sprouting of latent religious identities, which the fear of stigmatization or discrimination previously prevented from appearing. In this sense, prayer rooms in the city seem to carry less weight, although they remain a point of reference for the old pioneers and new arrivals alike, serving not only the typical purposes of religious organizations in emigration but also—and primarily—an identity rather than a merely religious function (Voas & Fleishmann, 2012; Rizzo et al., 2020; Groppi, 2020).

On the contrary, for the second group religion is a key element of identity, sometimes even in contrast with their parents' generation, which has developed a more private, less visible, religiosity.

> My mother does not wear the veil. I decided to put it on after a trip to Egypt. Even though we were born in Italy, we cannot deny our roots. And religion is part of those roots. I am not afraid of saying that I come from a country rich in culture, important in Mediterranean history. I'm proud to be the daughter of Egyptians, proud to be Muslim. My mother has made a different choice. She has stopped struggling. We know that life is not so easy for Muslims here in Italy. Today it is a little different: many of us now wear the veil at university, and nobody makes smart cracks or looks askance when we go round and about, to the cinema, shops and pizzerias. Twenty years ago it was different so, to cut a long story short, my mother stopped wearing the veil so she no longer has to be a target (F, Egyptian, 25, Turin).

Compared with a few years ago, the girls whose behaviour seems to be an interesting weathercock in the complex reality (including religious) of the children of immigration are more visible (Fassari & Pompili, 2019). Their adherence is convinced, visible, proud and active: being Muslim is bound to associational involvement where the religious theme joins those of social cohesion, discrimination and citizenship.

Those who are involved with religious associations are aware of the distance separating their generation and their parents with regard to living and interpreting their faith as well as relations with Italy and their country of origin. Intergenerational comparison of religious aspects shows up a deep reflective capacity on the part of the young, especially those with a higher level of education, in understanding the challenges facing them as children of immigration.

> Your idea of a Muslim is someone who is always going to the mosque, who does only what the imam says, who observes Ramadan. To me and many of my friends, being Muslim means coming from a family, which is attached to Islam. By now many of us young people only observe Ramadan and take part in feast-days such as that of the Ram. We are Muslims in our own way. We live here, not in Morocco or Egypt, so we should try to adapt (M, Moroccan, 24, Turin).

They also perceive the differences with regard to their parents' education and socialisation, which took place in environments permeated with religion where cultural, religious and national belonging was forged together into a unique affiliation—without distinctions within the local community, distinctive vis-à-vis interaction with the world outside.

In every migratory experience, at the generation shift, parents' associationism faces up to that of the young, whom they would like at their side to give new energy to their activities – without, however, letting go of the reins of command or calling into question the philosophy or objectives of the associations. *Is this true also for Muslim associationism?*

An initial point of difference concerns the characteristics of the associationism. Among young people, it is a matter of a reflection and a commitment path transversal to their origins: the criterion of access is that of recognizing Islam as their cultural-religious point of reference. As one of them recalled, "We don't ask our members for a certificate testifying that they are good Muslims. Our association is called Young Muslims of Italy, so the access criteria are clear. We are not bound to any particular country. Italy is our common reference and that of our section is Turin" (M, Moroccan, 25, Turin).

Another point of difference attains to leadership. Young people choose election-type mechanisms whereas their parents relied on consensus: in one case, we find elections, directives, pre-established deadlines; in the other reputation (as a good Muslim above all, but also endowed with elevated cultural and social capital) is the determining criterion in choosing.

A third element of difference has to do with the gender component: girls are an active part (although they have not yet reached the presidency), sometimes leading in organizing activities; mothers, on the other hand, keep a low profile. Here, we see not only a generational but also a gender revolution (Fassari & Pompili, 2017; Göle, 2014).

The last point deals with the local administration's demands.

What positions, then, do young people belonging to Islam adopt when facing the demands made on religious building, inter-religious activities, and involvement in lay-civic initiatives? Do they stand alongside their parents or do they keep their distance by offering their own vision/interpretation of the questions under debate?

The mosque question may serve as a prism dividing the positions of first and second generations. As a young interviewee reminded us, the mosque represents—in a context of mature immigration, for firmly settled backgrounds—a request that may no longer be postponed:

> There is no religious life here. It doesn't exist. If I am religious and want to practise, I can't. Italian mosques are the ugliest places in the world. This is one of the main reasons why I want to go back to Egypt. Here, I can't practise the way I would like to: there are neither instruments nor structures. Relationship with religion is difficult here because you are in a different society. Islam is a religion for the whole world but if you have no mosque, you suffer because it's hot and smelly with people shamefully packed together, if they close the mosques it will be extremely hard to practise one's religion and develop one's religious ideas (M, Egyptian, 26, Turin).

Parents and children are in agreement about this demand, but with a different approach and attributing a different meaning to it. For the latter, a mosque is now only a religious point of reference, and it should be considered as such in its structure and in its décor. For the men, above all, it is also for "recovering status". Parents, seeing their authority under threat as their children rush into integration and social insertion, try "to recover status as members of the mosque and find the symbolic motivational strength to transfer it within the family" (Dassetto, 1994, p. 73). This different approach implies a logical evolution from one generation to the next: all the younger interviewees reduce the mosque to a mere religious function, thereby creating clear discontinuity with the first generation.

For the parents' generation, the cognitive framework within which this relationship is set is that of immigration, which levers the dialectic between a community whose cultural-axiological roots are sunk deep in various *'elsewheres'*, but they are one as to religious reference and a hostile environment with which they are having trouble communicating. Their children would like to drop the references to immigration and diversity: the game is played among equals, between (almost) citizens, residents committed to the common good of the collective and the town or city. The change of tone is meaningful: as various scholars claim within the framework of the multi-level governance of integration policies, the passage is from being destined to be the object of interventions to being fellow-actors in developing policies (Meer et al., 2016; Keast et al., 2014). In this sense, the intercultural variation, in the meaning (which we shall see in the following section) of "inclusive intercultural policies", drives relations between Islamic associationism and local institutions to abandon the explosive, reductive immigrants-versus-citizens dichotomy. What is more, for second generations, it is no longer a matter of forwarding demands that limit Islam to a question regarding immigrants looking back nostalgically towards the past but of inserting the religious debate into the broader discourse of religious pluralism, unchaining it from its nexus with immigration.

Energy is spent on constructing relations of partnership, on gaining credibility and recognition: in other words, becoming trustworthy interlocutors of institutions and schools. Indeed, often the new actors have no experience of migration, and are often Italian citizens, which is to say, young people of another religion trying to find space, who express the desire to participate in their city's intercultural and interreligious politics.

References

Aluffi, R., Beck-Peccoz, R., & Zincone, G. (Eds.). (2008). *The legal treatment of the Islamic minorities in Europe*. Peeters.

Ambrosini, M. (2014). Migration and transnational commitment: Some evidence from the Italian case. *Journal of Ethnic and Migration Studies, 40*(4), 619–637.

Anselmi, M., & Blokker, P. (Eds.). (2019). *Multiple populisms: Italy as democracy's mirror*. Routledge.

Armfield, G. G., & Holbert, R. L. (2003). The relationship between religiosity and Internet use. *Journal of Media and Religion, 2*(3), 129–144.

Barzilai-Nahon, K., & Barzilai, G. (2005). Cultured technology: The internet and religious fundamentalism. *The Information Society, 21*(1), 25–40.

Bastenier, A., & Dassetto, F. (1993). *Immigration et espace public*. L'Harmattan.

Birnbaum, M. (2015). *Why is tiny Belgium Europe's jihad-recruiting hub?* The Washington Post, 17 January.

Bonino, S. & Ricucci, R. (eds) (2021). *Islma and Security in the West*. Palgrave.

Bonjour, S., Rea, A., & Jacobs, D. (Eds.). (2011). *The others in Europe*. Éditions de l'université de Bruxelles.

Bouzar, D. (2015). *Comment sortir de l'emprise "djihadiste"?*. Les éditions de l'atelier.

Cadge, W., & Ecklund, E. H. (2006). Religious service attendance among immigrants: Evidence from the new immigrant survey-pilot. *American Behavioral Scientist, 49*(11), 1574–1595.

CLIP. (2010). *Intercultural policies in European cities*. European Foundation for the Improvement of Living and Working Conditions.

Conti, B. (2018). Islam in Italy: From community to citizenship. *Oasis, XIV*(28), 54–69.

Crul, M., Schneider, J., & Lelie, F. (2012). *Second generations in Europe*. Amsterdam University Press.

Dassetto, F. (1994). *L'Islam in Europa*. Edizioni Fondazione Agnelli.

Fassari, L. G., & Pompili, G. (2017). On the learning process to be Italian Muslim women. *Scuola Democratica, 8*(3), 589–606.

Fassari, L. G., & Pompili, G. (2019). Performance as embodied narratives in young Italian Muslim women. *Italian Journal of Sociology of Education, 11*(2), 109–125.

Gans, H. J. (1994). Symbolic ethnicity and symbolic religiosity: Towards a comparison of ethnic and religious acculturation. *Ethnic and Racial Studies, 17*(4), 577–592.

Göle, N. (Ed.). (2014). *Islam and public controversy in Europe*. Ashgate.

Granata, A. (2010). Di padre in figlio, di figlio in padre. Il ruolo innovativo delle seconde generazioni nelle comunità religiose di minoranza. *Mondi migranti, 3*, 86–100.

Groppi, M. (2020). Islamist radicalisation in Italy: Just A myth? *Journal of Policing, Intelligence and Counter Terrorism, 15*(2), 117–135.

Hegghammer, T. (2010). The rise of Muslim Foreign fighters: Islam and the globalization of Jihad. *International Security, 35*(3), 53–94.

Keast, R., Mandell, M. P., & Agranoff, R. (Eds.). (2014). *Network theory in the public sector. Building new theoretical frameworks*. Routledge.

Kmec, V. (2018). Religion in the context of migration and youth in Europe. In A. Schnabel, M. Reddig, & H. Winkel (Eds.), *Religion im kontext | Religion in context. Handbuch für Wissenschaft und Studium* (pp. 253–268). Baden-Baden.

Koehler, D. (2015). Contrast societies. Radical social movements and their relationships with their target societies. A theoretical model. *Behavioral Sciences of Terrorism and Political Aggression, 7*(1), 18–34.

Leonini, L., & Rebughini, P. (2010). *Legami di nuova generazione. Relazioni familiari e pratiche di consumo tra i giovani discendenti di migranti*. Il Mulino.

Levitt, P. (2004). Redefining the boundaries of belonging: The institutional character of transnational religious life. *Sociology of Religion, 65*(1), 1–18.

Meer, N., Modood, T., & Zapata-Barrero, R. (2016). *A plural century: Situating interculturalism and multiculturalism*. Edinburgh University Press.

Monti, P. (2017). From social practices to reflective agency: A postsecular ethics of citizenship. In D. Thunder (Ed.), *The ethics of citizenship in the 21st century* (pp. 127–144). Springer.

Pace, V. (2004). *L'islam in Europa. Modelli d'integrazione*. Carocci.

Pew Research Centre. (2017). *Europe's Muslim growing population*. Pew Research Center.

Pless, A., Tromp, P., & Houtman, D. (2020). The "new" cultural cleavage in Western Europe: A coalescence of religious and secular value divides? *Politics and Religion*. https://doi.org/10.1017/S175504831900049X

Prinsen, F., De Haan, M. J., & Leander, D. (2015). Networked identity. How immigrant youth employ online identity resources. *Young, 23*(1), 19–38.

Ricucci, R. (2014). *Second generations on the move in Italy*. Lexington Books.

Ricucci, R. (2017). *Diversi dall'Islam. Figli dell'immigrazione e altre fedi*. Il Mulino.

Rizzo, M., Miglietta, A., Gattino, S., & Fedi, A. (2020). I feel Moroccan, I feel Italian, and I feel Muslim: Second generation Moroccans and identity negotiation between religion and community belonging. *International Journal of Intercultural Relations, 77*, 151–159.

Roy, O. (2002). *Globalized Islam. The search for a new Ummah*. Hurst.

Roy, O. (2007). *Secularism confronts Islam*. Columbia University Press.

Saint-Blancat, S. (1997). *L'islam de la diaspora*. Bayard.

Saunders, J. B., Fiddian-Qasmiyeh, E., & Snyder, S. (Eds.). (2016). *Intersections of religion and migration*. Palgrave Macmillan.

Scholten, P. (2011). *Framing immigrant integration: Dutch research-policy dialogues in comparative perspective*. Amsterdam University Press.

Silverstone, R. (2005). Mediation and communication. In C. Calhoun, C. Rojek, & B. Turner (Eds.), *The international handbook of sociology* (pp. 188–207). Sage.

Thomson, D. (2014). *Les Français jihadistes*. Les Arènes.

Van Tubergen, F., & Sindradóttir, J. (2011). The religiosity of immigrants in Europe: A cross-national study. *Journal for the Scientific Study of Religion, 50*(2), 272–288.

Voas, D., & Fleishmann, F. (2012). Islam moves west: Religious change in the first and second generations. *Annual Review of Sociology, 38*(1), 525–545.

Chapter 5
The Intertwining Identities of Being Both Children of Migrants and Catholics

The spectre of the "Polish plumber" has been around for a long time in Europe. He is the figure par excellence of the immigrant who is willing to do anything, a dangerous competitor who can change the dynamics of the labour market. Only the most attentive realised that in order to read the profound social changes brought about by the migratory flows, it was necessary to take care of (and worry about), and perhaps above all, the women: Polish, but also Romanian and then Ukrainian, Moldovan and so on. Not to mention other migrants who had arrived from distant countries, in Asia and South America, who had already been part of the life of German, French, Italian or Spanish families. Invisible figures, "segregated" in care work to be carried out between the walls of their employers' homes, where they care for children and the elderly.

These are women workers who enjoy positive discrimination: they care for fragile groups in the host societies, giving other women the opportunity to devote themselves to their careers and personal fulfilment. They go unnoticed because of their cultural-religious camouflage. Catholic or Orthodox, it does not matter, they are part of the "positive" imagery that concerns Europeans, Latin Americans or Filipinos with a faith that recalls a common Christian root. On the migratory scene in Europe, which discusses post-secularization, advancing "nones", secularism and spirituality, religion continues to be an important factor of acceptance and integration, although it is subordinate and segregated in some professional niches.

If it is true that work and family reunification are among the main drivers of human mobility, how one fits into the host society depends on the so-called structures of opportunity, including religious capital (Pisarevskaya et al., 2020). In a Europe where religion continues to be present and where the relationship with Islam is often seen only in terms of security, those with a Christian-Catholic background have an important advantage in terms of reception.

As we have already seen, parishes can be considered places offering refuge, respect and resources (Hirschman, 2004). However, the way in which these elements are used and perceived represents the real distinction between being accepted or stigmatized in the host society. For migrants, participation in places of worship

R. Ricucci, *Ethnicity, Identity and Faith in the Current Migratory Crisis*,
https://doi.org/10.1007/978-3-030-84056-3_5

has meant reconnecting and building new functional relationships in order to become autonomous, reliable and helpful subjects for newly arrived fellow countrymen (Ambrosini, 2016). Only at this point, migrants showing a low profile (e.g. not mentioning educational diplomas, job experiences, peculiar skills) were required to be successful in the search for a job, overcoming stereotypes, at least at some specific moments, to show their identity, including religion, in public (Brettell & Hollifield, 2014).

On the other hand, these aspects, for the second generations, highlight their limitations, as we try to investigate deeper. Religious environments are not for young people, the places where they "feel at home": whether born and brought up there by their parents or thanks to the intervention of religious figures who came from the mother country, from which at other times funding and requests also arrived, they are a clear reminder of the identity of an immigrant, an identity that interviewees, especially those born in Italy, escape from, not being an experience in which they were directly involved.

> Being a Catholic and a believer is part of my Filipino identity. Now I live in Europe and here religion is lived in a different way. Religion is not compulsory at school, there is no Mass at the beginning of the year and if you talk about religion, they tell you that you are 'strange'. There are not even processions where everyone participates, a party for the neighbourhood, the city. You realize that you are different just during the processions, which are organized outside the Philippines: you have the impression of being out of time (F, Philippines, 19).

> As children, it was fun to take part in the processions. It was a great party and there was a competition for those who earned and played important roles, either near the priest or in bringing something. Then, as you grow up, you look at everything with different eyes. They are the eyes of your schoolmates, of the people who stop to watch, to photograph. That's where you realize that you don't want to be part of that group. It's time to change, to become part of the environment in which we live. The processions are part of a story that is not ours; it is that of our parents and grandparents (F, Peruvian, 26, Turin).

An excellent example of this articulated social reality is represented by the religious processions of the migrants, who refer to traditions rooted in the countries of origin and bring elements of sacredness to the (secularised) European public space (Knott, 2005).[1] These are exceptional moments, which are organized through months of preparation in the smallest detail, with broad participation. From those who sew the costumes to the law enforcement officers, from those who sponsor the procession to those who prepare the prayers and songs. "Community" work, in which the protagonists are above all the first generations, who meet at and acknowledge an event that connects them ideally with the motherland and/or with relatives and friends who have emigrated to other countries or continents. The young people are called or, to use the words of an interviewee, "…receive the strict order to participate. The older ones are no longer interested in preparation, in which children

[1] It is a public space that over time has been enriched with characters, due to the arrival of other communities that requested its use for religious events. If, in a more or less easy way, everyone manages to navigate the red tape and obtain the necessary permits, it is mainly the Filipinos who are particularly benevolently regarded by the local Church and the Roman Church.

and adolescents are trapped. At the time of the party, however, everyone arrives. It is also a time for the family, not just the community" (M, Philippines, 25, Turin).

The procession becomes a collective happening: in fact, it also represents a moment of community effervescence, in which, in addition to faith, ethnicities are strengthened, where people meet as an ethnic-national group. At the same time, these events attempt to reaffirm the right to express one's own religiosity in the context of life outside one's home country (Corpuz-Brock 1996). In addition, processions can be counted among the factors, which migrants bring with them, displaying possible effects in host societies where the wind of secularization still blows together with other phenomena of religious revitalization (Table 5.1).

Ceremonies thus express both a secular value, which is a manifestation of identity, and a religious practice, which ideally reconnects migrants with their homeland through customs, songs and prayers. Numerous authors have described what similarly happened in Argentina and Australia with Sicilian, Venetian and Calabrian groups who emigrated from Italy. The same script is repeated at different latitudes and with different origins. Once, it was St. Rosalia and St. Rocco, statues carried through the streets of Buenos Aires or Perth by groups of the faithful. Today they are Santa Cruz in Barcelona, the feast of Los Milagros in Milan and the Black Madonna in London. In all cases, members of the chaplaincies (adults and young people, men and women) are involved in the months of preparation. Every detail is treated with the utmost precision because the religious event is also a public event, both locally and globally; at local level, because it represents a moment of affirmation of identity towards the host society; at global level, or rather transnational level, it is a moment of connection with the parishes of the countries of origin, with which many maintain contact through the remittances sent to support them or, as experiences of Latinos living in the south of the USA testify, with real dynamics of circular migration at certain times of the religious calendar.

Table 5.1 Immigrants' effects on the secularization process

Secularization means	The wind of secularization is changing its direction due to...	Emerging of new challenges
Vocations crisis	Increasing visibility of religions of immigrants	Managing religious diversity in the city: Muslims, different Christian groups, Jews, Sikhs
Decline in the number of rites' attendees	Widespread ethno-cultural religious affiliation	Accommodation of diversity in everyday school issues
Church's waning influence in the ethical area	Moving away from religious institutions to individuals	Political life: Tolerance of intolerant discourses in political life; public policies of exclusion; recognition of or opposition to minority mobilisation
Fading popular religiousness	… its presence among old people	The resurgence of street processions and confraternities

Source: Ricucci, 2014, p. 105.

The relationship with the homeland is also nourished through religion. Even with territorial differences and considering that in the places of the Diaspora, the whole territory cannot be involved, nor do its population sub-groups – the clothing, the sounds, the uniforms of the confraternity and its role remain identical. The people celebrate and pray in symbiosis with what happens miles away. In the era of ICTs, everything is transmitted and immediately shared, activating known mechanisms in the relations between emigrants and motherland.

The success of the migratory project is also evident in the magnificence of the procession, in the costumes used for the occasion, in the personalities involved: in a mix of sacred and profane, consuls, local administrators and successful entrepreneurs appear next to the priests. Then the donations are collected; significant in their amount, they are a tangible symbol of the economic power gained abroad, portrayed in photos and videos sent to relatives in the country (Garbin, 2019).

Once again, the Italian case qualifies as emblematic for the magnitude of the migration phenomenon and its diffusion in the territory and in society, and can be used as an explanation of the continental reality. For many Catholic migrants, arriving in Italy has meant achieving a double objective: fulfilling their project as economic migrants and living in a country where they can maintain their values. A union that quickly turned out to be a resource from a professional and work point of view, to be submitted to potential clients and employers and transformed into a competitive advantage. Women, often with a medium or medium-high level of education, with professional qualifications in the social and health field, and Christians (and therefore considered close in culture to Italian families) have been able to overcome the competition of other foreigners ready to enter the market of domestic care and assistance.

5.1 Catholic Diasporic Style Close to the Pope

A diachronic analysis of the data on migration flows in Italy underlines how the origin of migrants has moved progressively first towards Latin America and more recently Eastern Europe, with migrations from Poland, Romania and Ukraine becoming more and more relevant from the late 1990s. Yet, despite these data, research is still focussed on Muslims while an increasing number of Catholics (and other Christians) are now visible on the migration scene (Casanova, 2009). Furthermore, within the subgroup of those who belong to Catholicism, the youth perspective and their religious point of view are also underrepresented in comparison with the attention devoted to youth with a Muslim background. Speaking about children of immigrants, relations with religion need to be framed in two different perspectives. On the one hand, there is the acculturation process, i.e. how they fit into the receiving society (Portes, 1996; Berry 1999); on the other hand, how they define their identity as young people with a migratory background in a globalized world (Featherstone 1995; France 2007). In the case of children of those migrants with a Catholic (or a wider Christian) background, the religious belonging could not be a cause of stigmatization. Indeed, as we point out below, the daily life could be surprising: discrimination and stigmatization seem to be a process that also those Catholics with a migratory background have to cope with.

The way in which immigrants' cultural identities change over time and under the influence of the larger society, as do the characteristics of the relationship among them, are generally crucial issues in the study of the integration process of second generations. For children of immigrants rejoining their parents who emigrated before them, this process is more complex because it overlaps the larger development task of identity formation, which started in their home countries and then continued in the receiving society. Outcomes of this process mirror the various possible acculturation attitudes. Adolescents and young people can choose among various options: if they do not solve the cultural identity issues that they face, they may exhibit identity diffusion (Marcia, 1994). They acquire within the family the adaptive patterns of behaviour, personal characteristics, values, and social responses expected of them in their heritage culture. This is why even the issues of religiousness, religious belonging and the way of being religious need to meet at the crossroads of various sociological (and psychological) disciplines. Indeed, as Berry (2006) demonstrated, there is clearly evidence of the importance of religion in how immigrants engage in both their intercultural relations and their integration paths. Moreover, specifically, Christians earn the highest integration score. What is happening in Italy?

There are many possible patterns in the relationship with religion, with variations that depend on three sets of factors: (a) Personal and demographic factors (e.g. gender, age, length of residence); (b) Factors that are external to individuals and their communities (policies towards diversity in society); and (c) Differences between immigrant groups and society at large (such as differences in values).

The children of immigration we are discussing have mostly had some contact with their communities' religious milieu, and all define themselves as being in some sense religious. We do not know much about their relationship with religion. How do they interact with/develop their religious identity? Are they followers of the "traditional religiosity" of their parents (the first generation) or do they belong to the "secularized millennium generation"? Are they looking for a different (from their Italian peers) relationship with religion, more spiritual, requiring regular meetings, or do they want to get away from this? Is this generation seeking religious IT tools and means of contact with the church, less visible and "non-inserted" in the integration process among peers? Alternatively, on the contrary, do they choose, strategically, to reinforce the Catholic part of their identity in order to succeed better in the integration process in a Catholic country?

Italy seems—again—to be a good explanatory case for answering these questions. Indeed, from the religious perspective, the Italian landscape has profoundly changed because of the considerable Eastern European migration flows. Romanians, Ukrainians and Poles have modified the religious composition of the migrant population. Although the media continue to stress the risk of Islamization (Allievi, 2012), it is incorrect to speak of an "Islamic invasion", primarily because of the statistics:

according to estimates by Caritas, Christians constitute 48% of immigrants, Muslims 37% and oriental religions 7%[2] (Idos-Unar 2020).

It is within this context that the chapter focuses on the three immigrant groups, which are representative of the great majority of the non-Muslim immigrant population in Italy as well as in Europe: Filipinos, Peruvians and Romanians, groups that enjoy a good reputation among Italians (Istituto Cattaneo, 2018). Filipinos—particularly women—were the first to arrive. Today, the community is numerous, with a high incidence of family reunion, adolescents and young people who received the call to Italy after years of being "children left-behind" (Barbiano di Belgiojoso & Ortensi, 2019; Kofman & Raghuram, 2015). On the other hand, the Peruvian community began its immigration flow towards the middle of the 1980s. This was also principally female: as with the Filipinos, South American women left first, to be joined later by their children and, possibly, their partners.

The Romanians present different characteristics: their migration exploded in the 2000s with a greater balance between the sexes, even if women were the majority (Anghel, 2015). In any case, the three groups share the same characteristics: (1) Relative ease of insertion into the labour market; (2) Prevalence of care giving among the women; (3) Positive acceptance on the part of Italian society; (4) A high rate of children left behind and rejoined years later with their families; (5) The positive image of the children in schools. Using an American concept, the Italian media, government and schools, as well as the public, portray Filipinos, Peruvians and Eastern Europeans as the "model minority", contending that ethnic cultural traits predispose them to be financially and educationally successful.

These images foster borders between these and other immigrant groups. Two phenomena upset this peaceful scene: the fact of young Latino gangs; and racism against Romanians when they are all equally considered as being of Romany origin. In any case, these two issues do not erode their positive perception among Italians (Fellini et al., 2018).

What relationship do the children of Catholic-Christian migration enjoy with religion? Do they perceive it as a bond with the old country's traditions or is it a deep reminder of belonging untouched by the migratory process? It is well known that migration affects religious belonging, either weakening or strengthening it: how religious identity and belonging are at risk among young foreigners inserted into a social context where Catholic-Christian values are not seriously questioned but rather make up an important element of cultural belonging.

For the parents' generation, as we have seen in the previous sections, religion has been a card to play in the dynamics of insertion in the labour market and acceptance: Filipinos and Peruvians have been welcomed, partly because of their Catholic origins; Romanians, whether Orthodox or Catholic, have also benefited from their unproblematic religious belonging, which does not demand alternatives (to the churches) specific places of prayer, special work timetables to fit in with prayer or

[2] Apart from estimates of religious belonging, we do not have specific data about first generations or their children.

diets and ad hoc teaching in schools. The reunited children's generation from these three backgrounds undoubtedly gain from their ethnic-religious identity, which does not draw attention to them – as happens with their Muslim peers. However, this fact does not make them representative examples of religious culture in their home countries, nor does it immunize them against the effects of the adaptation process implicit in the integration path, including progressive distancing from institutional religion.

5.2 Your Religion Will Not Save You. The Uncancelled Immigrant Status

The children of immigration too are children of their age from the religious point of view and are therefore fully part of two dynamics the individual-religion relationship is undergoing. The first concerns the persisting belonging of the youngest to their parents' religion, in spite of growing religious pluralism and the advance of the "nones". In a broader sense, the Italian situation reveals particular characteristics compared with other European countries: higher average rate of religiosity; scant presence of other denominations; renewed presence of the Church in the public sphere.

Contrary to the expectations of many observers of modernity, we are not witnessing today a falling off from the religious sphere but a different way of belonging to it and expressing it (Stoltz et al., 2016). Furthermore, it is in this direction that we find the other dynamic, that of the growth of the individualism of faith. In the field of studies of religiosity, the decline in young people's attendance at religious services is generally considered as an indicator of the process of individualization of religious faith. Indeed faith is more and more a private fact accompanied by a reduction of participation in parochial life and observance of the religious intuitions' teaching on matters of moral and sexual orientation (Driezen et al. 2020). This aspect, along with the religious institutions' difficulty in reading and interpreting the needs of orientation and their answers to advanced modernity, makes Berger's "solitary believer" (1994) topical today, reinforcing what Davie identified as "believing without belonging". To these two dynamics can be added—for the "1.5″ generation—that of migration: the processes of individualization and personalization of living religion meet a community parenthesis in the initial period after arrival in Italy.

> A few days after I arrived in Italy, my mother took me to the nuns and one of them spoke Spanish. She welcomed me, showed me the recreation centre and told me that they organized activities for young people like me after Sunday Mass. I felt at ease. Before we left, we went to the chapel and prayed together. It was the first time I had done that for many years (F, Peruvian, 24, Turin).

> My mother took me to the church on Sunday. The first days I had walked around the city with her, feeling happy and overwhelmed. Although Italian is not difficult, it seemed like another planet. At the Romanian church, on the other hand, it was like going back. From

that point on I never missed a Sunday or any activity organized for us young people (F, Romanian, 24, Rome).

I have to admit that when I arrived I attended more faithfully. For us Filipinos religion is important – it is part of our culture. We grow up with Catholic values and they are what our parents teach us. Since I arrived, however, it has been harder: I have to work many weekends so cannot go to our Mass, Father Benna's—he says Mass in Tagalog and understands us. For those of us Filipinos who were born there and now are destined to live in Italy, it is important to have him as our priest. He's Italian and a bit Filipino. He understands us (M, Philippines, 25, Turin).

Several interviewees said they were Catholic or Orthodox believers: only a few Romanians said they believed in a superior being, without going into detail. In describing how they lived their lives as believers, all of them, more or less directly, referred to their ethnic churches, which, since their arrival in Italy, had been a point of reference. How this point of reference functioned, however, changed in time: if at the beginning it was the place of worship par excellence, where one reinforced one's identity, it later became principally a point of reference with regard to the cultural and linguistic environment.

Why do I come to this church? Well, it's not strictly for religious reasons, but cultural. Part of the reason is to keep some contact with the Filipino community because it's our only source of cultural identity. It's also important to our parents—this is not the most important factor, but one of the important factors. Another thing is that we think about the next generation. Our kids will have the same exposure if we have contact with the Philippine community (F, Philippines, 24, Turin).

According to another attending member, "The first thing that attracts people is fellowship, a sense of closeness. It's trying to find a group that's at ease. There is a sense of comfort in being with other Filipinos because there's an understanding in terms of background" (M, Philippines, 25, Milan).

It's like being part of an extended family. You feel protected, safe. You don't have to justify who you are, what you do, why you came to Italy (F, Philippines, 24, Milan).

I feel well when I am in church. I think of my grandmother at home and get a little homesick. She and I often went to church together. Here I can come with my mother only sometimes because she often has to work on Sundays, too. But, when we come we meet a lot of people we know and chat with them until it closes. It's like being in Romania and it does us good, especially when outside, round and about, we hear Italians complaining about foreigners, therefore about us (F, Romanian, 23, Turin).

For us Peruvians, Mass is an important occasion for meeting the community. It is much more than prayer: it is gathering as a family, feeling at home. Our parents feel as though they are back in Peru, not least because they speak Spanish, which is also the language of the Mass, so everything is simpler. The feeling for us young people is of turning back to the world where we grew up but don't live there any longer (M, Peruvian, 23, Milan).

Religion is left in the background: the reunion of compatriots, speaking the same language, the reassuring feeling that they are in a protective cocoon—this is what young people experience. This is not surprising when we consider that they have spent part of their lives elsewhere and at a certain point came to Italy, a context where being a child of immigrants—a foreigner—is not easy (Ricucci, 2012), where the insertion process is interwoven with the identity-building process of their age-group. This explains why attendance at ethnic churches is most frequent in the early years in Italy, and diminishes thereafter. Integration alone does not suffice to explain this divergence: it is mixed up with how these young people imagine the host society perceives them. For them too—Christian Catholics in a Catholic country par excellence—there are problems: paradoxically, they too have to deal with prejudice like their Muslim peers.

Race, an involuntary and ascriptive marker, is considered in the case of Filipinos and Peruvians to be one of the central features of identity, which separates them from the Italian society. Among Romanians, their association with both Romany and criminals risks negatively affecting their integration process: in these cases, both the colour of their skin and religious belonging can be useful to avoid being noticed.

However, one of the most significant ways in which young members of the churches articulate their ethnic identity is through the appropriation of certain elements of "practised culture", that is, values and standards of their traditional morality. These values, ubiquitously invoked in their discourse about their identity, consist of a set of core traditional values—most significantly filial piety, respect for parents, family-centeredness and a work ethic.

> My Philippine values include respect for elders, and emphasis on education. Another thing I love about my Philippine culture is its family-orientedness. I adore my parents. And I really like how children take care of their parents (F, Philippines, 24, Milan).

Compared to non-churchgoers, church members display a considerably higher degree of attachment to these values and standards of morality, which are appropriated by them as highly relevant criteria of their ethnic identity and as self-defined markers for signalling membership and exclusion. Indeed, the more "religious" members are, the more embracing they tend to be of the traditional values espoused by their parents.

5.3 Religion, an Element of Transnational and Intergenerational Identity

However, personal qualities and positive stereotypes alone were not sufficient to justify the construction of a positive image of Catholic communities: Catholic associations and ethnic chaplaincies themselves, which acted as mediating institutions (guaranteeing not only skills, but also moral rectitude and religious capital) towards Italian families for promoting positive images of their believers: the idea behind it is "They are true believers, so they are good workers. Please hire them" (Ricucci, 2014). A function that becomes part of family history, which some mothers have told their daughters in a perspective of religious socialization *sui generis*, as one interviewee remembers:

> My mother left our homeland after my aunt. Everyone in the Philippines, she says, had someone abroad [...] She didn't have much free time, but when she could she went to church to meet the priest, to meet other Filipinos. She says that she thought that in Italy living as a Catholic as in the Philippines would be easy, but it wasn't. She said that leaving work to go to Mass was not easy and often she had to give up. She said that the family in which she worked rarely went to Mass and they treated their elderly parents badly, yet it was a well-known Catholic family. She told us all the things they did against being Maka-Diyos [being for God]: it was instructive; they wanted to tell us what not to do and warn us against the risk of losing our identity as Filipinos and Catholics. Or perhaps Catholics first, and then Filipinos." And again, "Every year we participate in a feast organized for all migrant Catholic communities. My mother says it's a thanksgiving: in fact, many associations of nuns or people close to the church have helped many Philippine women to find work and have recommended them for their qualities. She said that in the nineties, finding work was not difficult, especially if the church recommended you. We Filipinos, she says, have the faith that unites us that other immigrant women do not have, not even Peruvians, for religion: many of them are Pentecostal and Italians do not like them. Then came the Romanians and the Ukrainians, they compete with us, also because Italians don't distinguish between Orthodox and non-Orthodox (F, Philippines, 24, Turin).

The Catholic identity of mothers has been a strategic resource for work. It has also been a resource to be strengthened within religious associations. In chaplaincies for Catholic migrants, social capital is also strengthened. A useful capital not only to orient oneself in the dynamics of integration processes, but also to be able to play the role of parents and guardians of the transmission of religious values. Being mothers and fathers in emigration brings new challenges, sometimes made difficult by the lack of a common language. Religious places and priests become allies in the parental task.

The theme of social control of the first generations over the second emerges transversally in many interviews, but it is particularly felt by young people of Philippine origin and by girls in particular.

Religious environments, in addition to the family nucleus, are institutionally important places for socialization and the maintenance of traditional values. Scientific literature considers the family as the central factor for the transmission of faith. In Italy, which is changing rapidly from a demographic, ethnic and religious point of view, what we know about this role in transmitting religious identity and practices can, however, be partially questioned.

The interviewees indicated mothers, rather than fathers, as people able to exert a strong influence on their religiosity, a figure that fully corresponded to the research carried out so far on the subject, not only in the field of migration. However, the meaning of this parental influence included an assortment of different practices not defined in the scientific literature on the relationship between religion and family.

On the other hand, the results of the research attributed greater importance to the role of ethnic congregations in migrant communities, highlighting the overlap in the influence of the extended family, and not only of parents, both at home and in religious organizations. In line with the findings of studies on religious transmission in the United States, young people pointed to other family members as key agents in their religious socialization (Alba, 2005).

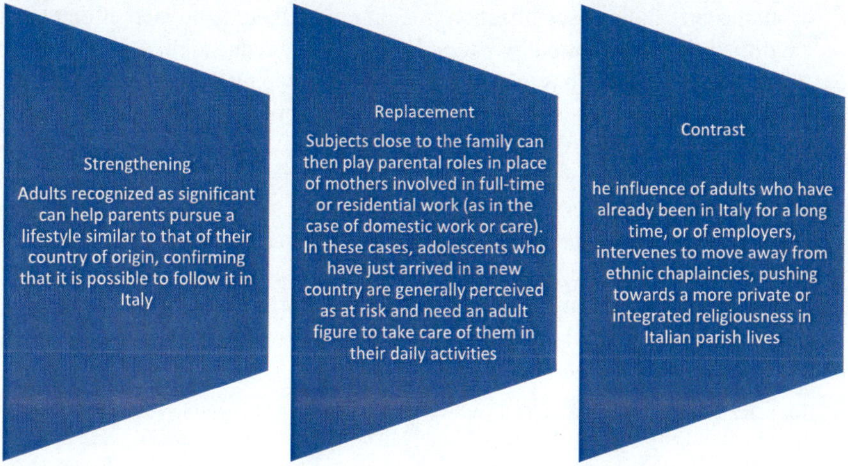

Fig. 5.1 Paths influencing youth's religiosity. Source: own elaboration

In particular, persons of reference who are not strictly part of the family are described as being able to influence religiosity in three ways, which we can conceptualize as "strengthening", "replacement" or "contrast" (Fig. 5.1).

Among the three options, the last one seems to be more feared by parents than the first, as the words of Minda, a 56-year-old Filipina mother, who arrived in Italy (Milan) back in the 1990s and waited 8 years before reunion with her daughter, shows:

> Young people here in Italy have too much freedom and for this reason they are uncontrollable, disobedient, have no respect for the elderly, for their parents. This would never happen with us. Here, no one helps you. At school, there is no discipline. My daughter tells me that there are companions who are on first-name terms with the teachers. With us, if you do not stand up when the professor arrives you are punished. If school doesn't help us, we can't do it on our own. Many young people already have histories that dishonour the family and make mothers ashamed. I pray that nothing like this will happen to my daughters.

I: Can you give us any examples?

> One girl was left by her boyfriend and she already had a baby, another went to live with a Moroccan boyfriend in another city. That's not why we left. We sacrificed ourselves to give our children a future, a respectable life to be presented to the community. In church, there are always mothers who ask for the help of the priest. If we do not rely on faith, who can save us here? We cannot go back, we must work. Our only strength is the Church.

I: There are public services, do you know about them?

> It's impossible for a Filipino to go and tell these things outside the community. We also have the task of not dishonouring our nation. We also remember it in church.

Minda is not the only one, as she herself pointed out, to turn to chaplaincies to resolve the intergenerational tensions that arise in immigrant families. Even more so when faced with a process of family reunification after years of distance.

In studies on religious socialization in emigration, there is no lack of emphasis on the difficulties encountered by parents in this type of education. Again, Italy, as a country of immigration, has the opportunity not only to read about what has been done, but also to translate successful experiences into practice and operational projects, avoiding those that failed. For many parents in emigration, religion is considered as a protective factor against the possible "corrupting agents" of the immigration society. Analysis of the educational offer addressed to the younger generations reveals, however, its distance from the feelings and characteristics of the children: often the activities are designed for the needs of the first generations, in which the latter are often involved only as recipients and not as organizers.

It is not surprising that they do not participate, or move away from each other, as another interviewee explains:

> Our parents haven't realized we're no different from them yet. We don't have to prove anything to the other Filipinos who are in Rome, Milan or other parts of Italy. We must grow up in this country and in this time and be with our peers. Religion is only a part of what we are and if we don't go to Mass it's not because we have become irreligious. We don't like the Mass designed for our parents, where in the end we distribute flyers with job offers or are invited to attend meetings about residence permits. None of my Italian friends goes to Mass or if they do, they don't say so. Nobody spends their time preparing food for the market afterwards. Our parents, I mean those of us of Philippine origin, demand that on Sunday we have lunch with them and the others in the community after Mass. It's okay if your boyfriend's there, but otherwise it's wasting your time. Look at all of us with our mobile phones in our hands, on Facebook or on Instagram, nobody talks to the others, everyone is waiting to leave.

I: Do you call yourself a believer?

> Of course, I am an Italian-Catholic Filipina and as such I can go to Mass wherever I want, in all the churches. Why do I have to go only to the church where Tagalog or Cebuano is spoken? Alternatively, where everyone knows me? So they can control me better? But then it doesn't concern my religious depth, only that I show myself to others (F, Philippines, 23, Turin).

> I'm a Catholic; I go to Mass, though not just to pray. It's a way to be together, to find myself with the other Filipinos. I don't go every week. Let's say once a month, because I know that my parents care about it, especially my mother. She says that the parish is the only place that helps us when we need it. She says it because it was like that for her when she arrived in Italy and today if a Filipino has a problem she goes to the priest, she doesn't go to a lawyer; to an accountant... she listens to what the priest says, prays and then leaves. Religion is part of being Filipino. For us young people a little less, but in Italy, too, it's like that. For my friends' grandparents going to Mass was an obligation not to be missed, today my friends don't even know who the Pope is (M, Philippines, 24, Turin).

> When you are a child, you like spending Sunday afternoons with other Filipino children, playing, singing, preparing the Christmas choir. But when you are about 15-16 years old, Sunday Mass starts to weigh you down. You know, Mass can last two hours, and it is partly in one of the Philippine languages. I don't know if you've ever been? At that age, you don't want to spend time praying and so you start to find excuses with your parents, stay in the

back of the church, leave as soon as you can. Our parents don't understand us. Therefore, they force us to attend activities for other Filipinos. It's just that sometimes you feel a bit ridiculous. For example, when we do the contest for Miss and Mr. Philippines, we do it in the church hall; it's a charity event, where our parents also invite their employers. It's a bit American. It's organized to get us involved, to keep us in the community. It has nothing to do with religion, but it begins with prayer and the priest is with us all the time (M, Philippines, 25, Milan).

5.4 The Great Illusion. We Don't Speak the Same Language, But Maybe We Believe in the Same God

The generational changeover is also captured through the language in which you pray, and one prefers to participate in the rites. The family, whether re-joined or constructed in exile, divided or single-parent, far or near, is a key element both in the migratory process and in analysing the paths of young generations. In the former case, it may serve as an indicator of, for example, stabilization, long-term migratory processes, of deferral, of the idea of return. In the latter, it represents the environment within which one grows and/or where one arrives, in which one learns behaviour and rules, but also where conflicts grow and people split up, more likely in families where both parents and children are involved in a parallel and interrelated process of change in self-definition, with ambivalent tensions between 'over here' and 'over there', towards the country of origin and that of residence. The distance between parents and children may grow as a result of school attendance, language acquisition, frequenting friends and the context of the host society: young people acquire cultural and linguistic expertise faster than their parents. The latter are "forced into silence": their home language is looked down upon and banned in the sense that it is considered an obstacle to the children's learning the language of the new country, a language that makes the parents keep quiet because they often know and use it badly (Chao, 2006; Kam & Lazarevic, 2014).

Moreover, it is precisely because of these missing skills that children often have to play the parts of translators and interpreters, even representing the family in the outside world as the family's ambassadors to Italian society. The parents' scant knowledge of the second language, together with the children's disparagement of the language of origin, brings about a situation where it is almost impossible to have a fully educational relationship within the family. In relationships where dialogue and discussion of personal matters are rendered difficult by the lack of a common language, parental support through the various phases of growth is limited because, on the one hand, the language of love and care and, on the other, the possibility of bonding through symbols, rites and words of the parents' home-country culture have been cancelled by the migratory experience (Weisskirch, 2013).

The feeling of loneliness is commonly expressed in situations of reunion between adolescents and mothers (Chib et al., 2014; Madianou, 2012), but is still released in many fewer cases than are documented by ethno-psychologists or transcultural psychology operators.

> I arrived in Turin when I was 14. My mother had left me behind in Peru, entrusting me to the care of my father and grandparents for 3 years, during which time I grew up immersed in the Peruvian environment and culture. Now, here I am in Turin, in a completely different setting. My arrival gave rise to fleeting happiness, but my mother's working timetable allowed me to see her only rarely - we were faraway neighbours. When I started school, I met only problems: the language, study materials, the environments – but by persevering, I got my high school diploma. I am now thinking about how to acquire citizenship and perfect my knowledge of the language, which should make everything easier. It is not true that we Peruvians have a language advantage [in Italy, with respect to other immigrants, ed.]: we rather run the risk of not understanding and not being understood. This happens to my parents, who think they understand everything because they consider Italian and Spanish to be "similar": but then they ask me to translate or to confirm what they have understood, which turns out most of the time to mean what they have misunderstood. It isn't easy. They live in Turin as if they lived in a Lima neighbourhood, meeting only Peruvians, going to a Mass for Peruvians said by a priest who speaks only Spanish. They say that by now they are integrated; but that isn't integration, it's segregation. Is it possible that they don't realise this? On the contrary, they want me too to mix only with Peruvians and to marry a Peruvian girl, so that there will be no problems. My mother says to me: 'Listen, if you marry an Italian or a Romanian or an Albanian girl, what will I do? Will we lose our traditions and our bonds?'

I: What would she say about a possible Muslim girlfriend?

"Unthinkable. A different religion, a different culture. For our parents, children are not Italian but extensions of themselves; we are expected to follow them. Only our bodies are in Italy while our hearts and minds should be in Peru. We are expected to speak Spanish, eat empanadas at least once a day and hang around with other Peruvians" (M, 20, Peruvian, Milan).

Mothers plan and put their children's arrival into practice (school registration, free-time language courses) believing that the impact is painless, uncomplicated: everything will work out – they think – for their children who speak a language similar to Italian. As one interviewee said: "All you need is a little discipline. We need to keep an eye on our children so they don't go off the straight and narrow" (Romanian mother).

However, communication problems and difficulty in exercising parental control seem to go hand in hand with "strict discipline", especially where there has been prolonged separation – despite attempts to make up for time lost because of distance, and an empty house caused by absence due to entire days working, by the purchase of consumer goods. In doing this, parishes and ethnic chaplaincies become key players. Parents turn to priests and other adults who attend religious places to be supported in their parental role in emigration. The religious environment plays the role of moralising agent and censor of behaviour not considered acceptable.

The mother or father's departure in time leaves scars, which are hard to heal, especially at an age such as adolescence, which is already full of changes and relational problems. Indeed, having been "children left behind" influences all subsequent intra-family relationships, in areas such as parental authority and the value and importance attributed to the institution of the family in both the present and the future.

> I don't have much to say to my parents. We don't get on. They want to tell me what to do, where to go and how to behave. They are thinking of an ideal son, Italian at school and Peruvian at home. That's not the way it is. At school I feel all right, except that I'm a foreigner. It's not like being in Peru, in my aunt's house, where I felt fine. That is my home, where I am a Peruvian

(M, Peruvian, 23, Turin).

The same—unwritten—rules of migration apply to family reunions: the immigrant presents himself/herself to relations and compatriots who have stayed at home as someone who has made it, without mentioning the difficulties encountered, their legal position, or their precarious employment or accommodation conditions (Taliani, 2015).

The children especially, trapped between two cultures, are engaged in a (re)definition of their relations with their parents.

There are four ways in which such relationships are built and evolve emotionally. The first is alienation, which derives from an unresolved generational conflict, accompanied by a multifaceted cultural conflict, depending on the migration timing of parents and children, the relationship kept up with one's country of origin and the investment made in the welcoming country in terms of relationships formed with natives and compatriots (Table 5.2).

While overcoming the language barrier does not always help with the cultural and/or generational divide, it may reinforce the parenting feeling – looking after one's parents as if they were children needing one's help because they have a weaker grasp of tools of interaction with Italian society.

The third is solitude or loneliness: whether reunited or born in Italy, these young people's experience is often characterized by the lack of parental figures, as the latter are heavily engaged in their work.

The last feeling is frustration deriving above all from disappointed expectations of the success that these children are driven to by their parents who have social redemption in mind, and the ascending mobility promised by straight-line assimilation.

In families where both societal and language divide exist the relations across generations are poor: there is little or no parental involvement in their children's

Table 5.2 Differences between parents and children: the emergence of four relationships

		Who in the family is able to read Italian and interact with natives?	
		Parents & children	Only children
Who in the family speaks Italian?	Parents & children	Connection between generations FRUSTRATION (1)	Societal divide (2) PARENTING
	Only children	Language divide (3) ALIENATION	Language & societal divide (4) LONELINESS

Source: Ricucci, 2021, p. 251

areas of life (schools, associations, peer groups) or any possibility to play the parental role effectively.

According to the juvenile population's current experience, neither membership nor identity is exclusive: being pluri-associated is the rule among the Millennials. In this way, young people born and/or grown up in Italy are all equally bound to the reality in which they live. They are young Millennials, so their activities are not only carried out within physical ethnic parishes through weekly meetings, training and discussion weekends, participating in demonstrations, congresses throughout Italy but also, in some cases, on the web.

5.5 Second-Generation Catholics Improve the Secularized Youth Millennials

Broadly speaking, there is more to the lives of these youth than mere religiosity and attachment to their home countries. Despite common elements of identification and cultural specificities, they have as much in common with their peers, both Italian and foreign, as among themselves. While there are many lines of demarcation within the category of "children of immigrants" with regard to class, gender, education and cultural divide, to name a few, there are also certain common attributes that make the category of "Christian" meaningful. However, to understand what is happening we must bear in mind that this group stands at the crossroads of three sociological areas of interest: their youth, their assimilation process and their ethnicity. Considering these three facets, their religious belonging as well as their level of religiousness assumes relevance on two planes: in their relations with both the host society and ethnic communities; and in their interaction with peers.

Some young people have played down their ethnicity and their religion and have sought to "pass" as non-Filipino, non-Peruvian or non-Romanians. In other words, they try to cross ethnic (and religious) boundaries for integration purposes. Sometimes it is very difficult: skin colour and somatic traits prevent them from hiding their ethnic origins. This is certainly the case for Filipinos and Peruvians, whose appearance and accent give them away. However, if Romanians are apparently advantaged in choosing an assimilation strategy, they share with peers the daily struggle with their families and ethnic communities.

In other cases, young immigrants choose the middle road: they go to church and show loyalty to family values, but without either commitment or conviction. In this case, religious belonging is perceived in a social rather than a religious sense. It is quite evident that for those young people belonging to Catholic congregations inserted in the Italian-Catholic context, with positive relations at school or at work, neither the parents' religion nor their religiousness is very important. They respect their parents' desire to preserve certain traditions, but do not personally speak of them with the same enthusiasm. Nevertheless, while their experience differs from the first generation, it is equally distinct from Religion Italian Style, which means a country – in contrast with other European Catholic countries—where Catholicism continues to be strong and widespread among the population. Various data and indicators confirm the presence of a lived Catholic subculture permeating both public and private life (Eccles, 2015). The second-generation youth in these Catholic congregations have charted a path that contains elements of collective expression (festivals, language, transnational ties) as well as personal agency (intense religious experience, youth organizations). Their involvement in an ethnic congregation connects them not only to families and their ethnic community here, but also to their peers in their home countries. In this case, we are dealing with young people who behave "transnationally", being able to cross (in)visible cultural boundaries, developing their new hyphenated identities such as Italian-Filipino and Italian-Peruvian. Romanians stand out due to their European Union belonging: their efforts are devoted towards being considered European citizens as much as Germans and French and not only children of immigrants, which in the public debate means being negatively considered and condemned to be considered as an unwelcome guest.

The situation of those who have not established a positive relationship with Italian society, who have not managed to integrate, is different. They have remained anchored primarily to their family and community, declaring that they assiduously attend their ethnic church and its functions. In these cases, however, attachment to their community of origin *seems* to be dictated rather by necessity than by choice. Indeed, the community, and consequently the ethnic church, are elements of support for those young people with a limited knowledge of Italian who have not developed a mixed network of friends, but who have made contact with compatriots and activities organized within their ethnic circle as the main contexts of their identity. As for the first generation, the church becomes a place of refuge in the storm of Italian society.

References

Alba, R. D. (2005). Bright versus blurred boundaries: Second-generation assimilation and exclusion in France, Germany, and the United States. *Ethnic and Racial Studies, 28*(1), 20–49.

Allievi, S. (2012). Reactive identities and Islamophobia: Muslim minorities and the challenge of religious pluralism in Europe. *Philosophy & Social Criticism, 38*, 379–387.

Ambrosini, M. (2016). Protected but separate: International immigrants in the Italian Catholic Church. In D. Pasura & M. Bivand Erdal (Eds.), *Migration, transnationalism and catholicism* (pp. 317–335). Palgrave Macmillan.

Anghel, R. G. (2015). *Romanians in Western Europe: Migration, status dilemmas, and transnational connections*. Lexington Books.

Barbiano di Belgiojoso, E., & Ortensi, L. E. (2019). Satisfied after all? Working trajectories and job satisfaction of foreign-born female domestic and care workers in Italy. *Journal of Ethnic and Migration Studies, 45*(13), 2527–2550.

Berry, J. W. (2006). Acculturative stress. In W. J. Lonner & R. S. Malpass (Eds.), *Psychology and culture* (pp. 211–215). Allyn and Bacon.

Brettell, C. B., & Hollifield, J. F. (2014). *Migration theory: Talking across disciplines*. Routledge.

Casanova, J. (2009). Immigration and the new religious pluralism: A European Union – United States comparison. In G. Brahm Levey & T. Modood (Eds.), *Secularism, religion and multicultural citizenship* (pp. 139–163). Cambridge University Press.

Corpuz-Brock, J. (1996). Gospel, Culture, and Filipina migrant workers. *International Review of Mission, 85*(336), 63–84.

Chib, A., Malik, S., Aricat, R. G., & Zubeidah Kadir, S. (2014). Migrant mothering and mobile phones: Negotiations of transnational identity. *Mobile Media & Communication, 2*(1), 73–93.

Driezen, A., Verschraegen, G. & Clycq, N. (2020). Religion and everyday cosmopolitanism among religious and non-religious urban youth. Current Sociology. doi:10.1177/0011392120932942.

Eccles, J. (2015). Religion Italian style: Continuities and changes in a catholic country. *Journal of Contemporary Religion, 30*(3), 546–548.

Featherstone, M. (1995). Undoing Culture: Globalization, Postmodernism and Identity. Sage.

Fellini, I., Fullin, G., & Quassoli, F. (2018). La percezione della discriminazione etnica sul lavoro. In ISTAT (Ed.), *Vite e percorsi di integrazione degli immigrati in Italia* (pp. 167–192). Istat.

Garbin, D. (2019). Sacred remittances: Money, migration and the moral economy of development in a transnational African church. *Journal of Ethnic and Migration Studies, 45*(11), 2045–2061.

Hirschman, C. (2004). The role of religion in the origins and adaptation of immigrant groups in the United States. *International Migration Review, 38*(3), 1206–1233.

Istituto Cattaneo. (2018). *L'immigrazione in Italia: tra realtà e percezione*. Retrieved 1 December, 2020, from https://www.cattaneo.org/2018/08/27/immigrazione-in-italia-tra-realta-e-percezione/

Kam, J. A., & Lazarevic, V. (2014). The stressful (and not so stressful) nature of language brokering: Identifying when brokering functions as a cultural stressor for Latino immigrant children in early adolescence. *Journal of Youth and Adolescence, 43*(12), 1994–2011.

Knott, K. (2005). *The location of religion. A spatial analysis*. Equinox.

Kofman, E., & Raghuram, P. (2015). *Gendered migrations and global social reproduction*. Palgrave Macmillan.

Madianou, M. (2012). Migration and the accentuated ambivalence of motherhood: The role of ICTs in Filipino transnational families. *Global Networks, 12*(3), 277–295.

Marcia, J. E. (1994). The empirical study of ego identity. In H. A. Bosma, T. L. G. Graafsma, H. D. Grotevant, & D. J. de Levita (Eds.), *Sage focus editions* (pp. 67–80). Sage Publications, Inc..

Pisarevskaya, A., Levy, N., Scholten, P., & Jansen, J. (2020). Mapping migration studies: An empirical analysis of the coming of age of a research field. *Migration Studies, 8*(3), 455–481.

Portes, A. (1996). *The new second generation*. Russell Sage Foundation.

Ricucci, R. (2012). Youth migration in Italy: A new working class? *Power and Education, 4*(2), 230–243.

Ricucci, R. (2014). *Second generations on the move in Italy*. Lexington Books.

Ricucci, R. (2021). Language, emotion, and the digital divide among immigrant families. Evidences from Italy. In M. Borges, S. Cancian, & L. Reeder (Eds.), *Languages of love: Emotions, gender, and migration since 1880* (pp. 238–257). University of Illinois Press.

Stoltz, G., Könemann, J., Schneuwly Purdie, M., Englberger, T., & Krüggeler, M. (2016). *(Un) Believing in modern society. Religion, spirituality, and religious-secular competition*. Routledge.

Taliani, S. (2015). Antropologia dell'infanzia e delle famiglie immigrate. *Rivista della Società Italiana di Antropologia Medica, 39-40*, 17–70.

Weisskirch, R. S. (2013). Family Relationships, Self-esteem, and Self-efficacy Among Language Brokering Mexican American Emerging Adults. *J Child Fam Stud 22*, 1147–115.

Chapter 6
Faith in God is Our Strength! Faith-Based Organizations Challenging 'Fortress Europe'

> The reality of these stories is the reason why we intend to practice welcoming and integration in the centre, remembering that human dignity and solidarity among people are at the centre of everything.
>
> Sergio Mattarella, President of Italy[1]

"Pope Francis: A Man of His Word". This is the title of Wim Wenders's 2018 documentary about the first 5 years of the Argentinian's pontificate.

Part of the charisma of this prelate, presented as a consistent witness to Catholic values, is due to his linguistic register: from the beginning, the colloquial terms used by Jorge Mario Bergoglio, rich in references to everyday life, were seen to be his distinguishing trait.

Commentators and analysts soon added further themes to his communicative style: the church of the poor for the poor; the necessity for internal renewal of ecclesial structures and their paying greater attention to the surrounding society and its transformations; tension in dealing with the last and the marginalised; and the duty of welcoming. One may say that these are traditional teachings, but they are presented in a very different way, not least because of the symbolic places where they are transmitted. An example is the island of Lesbos, sadly famous because of the images of a refugee child dead on the shoreline—images too soon forgotten. It was the perfect setting to remind us of opening up to others. Then there was the stadium in Panama, chosen to invite young people, adults and the elderly to bear witness to the Gospel while remaining open to diversity (including religious) and otherness (counting foreigners).

Pope Francis's position with regard to immigrants was clear, his involvement not fully appreciated. A recent Italian survey allows us to examine this aspect of papal commitment among Italians. Only 66.7% were favourable, an indication of how thorny and difficult it is to deal with the reality of immigration, partly because of its physiological complexity (by the way, almost 19% of respondents answered "Don't know, no answer"). Almost 73% of affiliates of other religions agreed with the Pope; it is natural to suppose that part of the explanation lies in the—past or present—condition of migrants, of many in this category. Among the most opposed

[1] Inaugural ceremony of a Jesuit Refugee reception centre in Rome, 02/04/19.

R. Ricucci, *Ethnicity, Identity and Faith in the Current Migratory Crisis*,
https://doi.org/10.1007/978-3-030-84056-3_6

Table 6.1 Applications and granting of protection status at first instance: 2019

	Applicants in 2019	Pending at end 2019	Refugee status	Subsidiary protection	Special protection	Rejection	Rejection rate (%)
Total	43,783	42,803	10,711	6.935	616	76,798	81

Table 6.2 Breakdown by country of origins of the total numbers. Five top countries by number of applicants

	Applicants in 2019	Refugee status	Subsidiary protection	Special protection	Rejection	Rejection rate (%)
Pakistan	7305	710	919	53	10,272	85
El Salvador	5520	700	897	5	1581	49
Peru	2445	119	24	1	1172	89
Ukraine	1775	105	3380	39	3586	88
Venezuela	1544	447	981	6	11	7

Source: Commissione nazionale per il diritto d'asilo, Ministry of Interior

were people with no religion (over 20% of the total) and, among Catholics, those who said they were so by birth and upbringing (about 18%) (Garelli 2020).

These are interesting findings, partly because they are in counter-tendency against the visibility and participation of Catholics in welcoming refugees. Especially in the country where the pope lives: indeed, in Italy the issue of asylum seekers represents a key topic both in the media and political debates, greatly stimulated by the far-right parties (Longo 2019; Georgi 2019) (Tables 6.1 and 6.2).

Organizations with religious roots are at the forefront with regard to the refugees just mentioned: the latter present their own juridical status and specific characteristics in the motivations behind their migratory itineraries, and often weakness in their integration paths in the receiving society.

Indeed, focussing on the issue of refugees and asylum seekers, their rapidly increasing number, the effort undertaken by some countries and the resistance of others to their reception has been a real test for inter-ethnic and inter-religious relations. Throughout Europe, especially in the countries bordering the Mediterranean, the difficulty of managing the phenomena, and the increase in deaths at sea, has underlined the weak points in the relationship between inclusive and exclusive societies. Among the countries involved, Italy is the context in which the issue has been most relevant, and the political repercussions were not long delayed: the latest elections revealed a growing preponderance of right-wing and populist parties, interpreted by many as a reaction of the population to the growing arrivals and to the linked fear of the dynamics of religious opposition. In the public debate, all new arrivals belong to Islam and the fear of undergoing an Islamic invasion re-emerged.

However, religion is not only important as part of the identity of migrants and refugees or a political argument in the debate. In fact, action to face the so-called refugee crisis was and still is a fundamental role played by international, national and local religious organizations in terms of advocacy and concrete projects and strategies.

As can be seen in Chap. 1, despite there being a rich body of literature on migrations in these times (Crawley and Skleparis 2018), studies investigating religion within immigrant communities, including its role and impact on asylum seekers and refugees, have been less extensive. However, evidence from some contemporary ethnic groups suggests that ethnic religion may play a strong role in the lives of those who fit into the heterogeneous group of migrants. The chapter summarizes the main religious issues that are currently guiding the debate in Europe dealing with faith and refugees and asylum seekers, considering how religion and the settlement paths of asylum seekers interact with and influence one another, both at community and individual level, and the role (if any) played in this relationship by ethnic-religious associations and centres already established in the host societies. As mentioned in previous chapters, Italy has become an interesting case study in the above-mentioned framework because of the significant number of policies, initiatives and projects started there and then activated based on a policy-sharing approach in other European countries.

6.1 Faith-Based Organizations: Leading Players in the Management of Social and Religious Cohesion

The churches are the most important quantitatively, and best organized, migrant organizations among several immigrant groups. Far more than other institutions, they create a public space of everyday relevance for the migrants, specifically in those areas where there are scarce resources for marginalized groups of migrants. In all cases, we can identify a similar pattern in their formation: the founding process started after the first wave of migration, which varied from the 1960s to the 1980s in different European countries. The establishment of churches was connected to three developments: first, a continuous growth of the immigrant population; second, the need to answer their religious requests; and third, the process of family reunion. However, religious actors do not only play a role in welcoming societies. Indeed, "*The preponderance of political and socioeconomic considerations in discourses on transmigration and Diaspora often glosses over the religious factors that stimulate and impact on such processes. As we are faced with "space-time" compression, and a growing interconnection of local-global spaces, contemporary migratory trends bring to the fore the crucial role, functions and import of religious symbolic systems in new geo-cultural contexts*" (Adogame 2007, p. 20).

Whether liked or disliked, "religious" actors, "*shape governance issues in a global world and awareness of their involvement, value and contribution is vital for their justice, peace, and reconciliation on a wide range of policy issues*" (Carrette and Miall 2012, p. 11). These actors recite in different ways. At macro level, religious institutions could intervene "politically" according to their statements and positions. At meso level, the numerous actors who are in charge of putting into practice values and missions, in operationalizing what documents and political positions state (e.g. dioceses and parishes, mosques and temples, churches and all faith-based organizations). Finally, at micro level, the focus is on the volunteers and those who have a salaried job in the above-mentioned organizations.

There is a general consensus that FBOs are growing in numbers and importance. This seems paradoxical in the "post-secular" international framework and relations.

There are a number of terms in use in the scholarly and policy-related literature to refer to non-state actors motivated by faith-linked concerns: religious NGOs, religious international (national or local) non-governmental organizations (RINGOs), religious communities, religious actors/groups, faith-based non-profit organizations, and faith-inspired organizations. The issue is questioned, as the table below tries to show by summarizing the main classifications, which agree that faith is one of the key component dimensions of such non-state entities, which can operate at various levels, including local, regional, national, and international (Hirschman 2014) (Table 6.3).

Clarke and Ware (2015) offered another method of identifying and classifying FBOs, which can be, (1) *Faith-based representative organizations or apex bodies*, which rule on doctrinal matters, govern the faithful and represent them through engagement with the state and other actors; (2) *Faith-based charitable or development organizations*, which mobilize the faithful in support of the poor and other social groups, and which fund or manage programmes that tackle poverty and social exclusion; (3) *Faith-based socio-political organizations*, which interpret and deploy faith as a political construction, organizing and mobilizing social groups based on faith identities but in pursuit of broader political goals; (4) *Faith-based missionary organizations*, which spread key faith messages by actively promoting faith and seeking converts to it, or by supporting and engaging with other faith communities based on key faith principles; and (5) *Faith-based radical, illegal, or terrorist organizations*, which promote radical or militant forms of faith identity, and/or engage in illegal practices based on faith beliefs (Hirschman 2014, pp. 11–12).

As always, every classification has its limits. In this case, some organizations may overlap various criteria. For example, the International Islamic Relief Organization (IIRO), which fits into "*two categories, i.e. faith-based charitable organization and faith-based socio-political association, emphasizing that many*

Table 6.3 Main description of FBOs in the scientific debate

Tadros (2017)	Wuthnow (2005)	Castelli and McCarty (1997) and Vidal (2001)
Faith-based and/or faith-inspired	Interfaith- or multifaith-based	Local congregations
Religiously-oriented mission statement	Support from a religious organization	Being founded by a religious institution
FBOs based on national religious congregations	Faith-based networks	Freestanding religious organizations, which are incorporated separately from congregations and national networks
Ministries of religious faiths	Ministries of religious faiths and Lay people	Ministries of religious faiths and lay people

Islamic charities in the Arab world are informally linked to opposition movements, such as the Muslim Brotherhood or al-Qaeda, which falls into the category of a faith-based radical organization".

In addition to the classifications, the participation of young people—whether as recipients or as volunteers in activities addressed to refugees or asylum-seekers—is a transversal element; and this is a novel situation.

6.2 Religious Associationism and Asylum-Seekers: The Importance of Having a Working Method

Over the last 10 years, changes brought about by globalization, immigrant settlements and increased ethnic, religious and linguistic diversity in Southern European countries such as Greece, Spain and Italy have reinforced the need for policies capable of promoting social inclusion and preventing conflict (Jeandesboz and Pallister-Wilkins 2016). On the one hand, mainly due to migration, we are witnessing an increase in the presence of heterogeneous cultural groups as part of the socio-economic fabric of the towns and cities while, on the other, there is an exploration of how interculturalism can help foster social inclusion (Carella et al. 2007). In recent years, more of a connection has been forged between interculturalism and policies for social inclusion, a development that has sometimes given rise to effective inclusion initiatives, and sometimes remained on a more superficial level. This is the case, for example, of intercultural festivals, events devoted to raising awareness of the cultural traditions present in a town or city without, however, necessarily pursuing greater social inclusion of these groups or fostering closer relationships and intercultural dynamics between natives and migrants.

The results have been precarious balances, greatly alleviated by voluntary organizations, between this sector and administrative bodies interested in promoting intercultural dialogue and/or conflict mediation (Georgiana and Agafiţei 2016). The Mediterranean crisis fermented by the Arab Spring and consolidated by the Syrian exodus has called into question the integration processes underway in many European towns, cities and countries.

The main reasons have been identified as the growing numbers and biographical characteristics of the players (often poorly educated, dark-skinned and Muslim), and the increased welcoming costs. The tragedies of Lampedusa, Ceuta and Lesbos have attracted public attention to the subject—attention that has also made clear the crucial role played by religious associations and charitable organizations.

This has resulted in public appeals for welcoming refugees and asylum-seekers by the most important religious leaders on a global level, with effects at local level; crisis management actions of the various Christian and Muslim religious organizations (humanitarian corridors, first aid assistance and first reception), and events and initiatives, especially at local level, to promote meetings of refugees and asylum-seekers and the host community.

The various religious-associationism realities are often a multifaceted container for activities addressed to refugees and asylum-seekers (Ambrosini 2015). Looking at initiatives planned and carried out in recent years one observes the ability to build structures transversally supported by three pillars.

Above all, *rigour* in defining and carrying out initiatives. Volunteers make valid proposals characterized by high qualitative and professional *standards*. However, the *rigour* is also seen in the capacity to refrain from jumping into any activity just to be seen to be doing something. What is important is not action but *savoir-faire*, not what one does but through what relations and with what objective. Primarily, not rigid activities and methodologies but modalities that take shape as a function of recipients and their needs. This introduces the second structural element: the flexibility of proposals, which is to say their suitability to be applied to the daily life and experience of their recipients, whether newly arrived or (young or adult) reunited with their families, unaccompanied minors or asylum-seekers. This explains why not all associations are engaged in the same activities and (e.g. language courses are a recurring theme) even if they are, they are not always carried out in the same way.

Operators' characteristics, timetables and ways of managing spaces are modified and transformed as a function of recipients' most pressing needs—examples are travel times, working hours and prayer necessities. Finally, *innovation* understood as a capacity for interaction between *leaders* and *followers* or, in this case, positive dialectics between operators and volunteers.

Further qualities are added to the pillars upon which every activity rests. They may be considered as guiding principles for everybody working in FBOs. The *spontaneity* with which one carries out one's tasks is typical of the passion for education (sometimes inspired by religious motives) upon which these organizations base their efforts. This spontaneity does not mean chaos, the lack of a common thread or an educational plan/pact with the recipient. A corollary of spontaneity is *informality*, meaning rather a friendly atmosphere than one without rules. Everybody benefits from this atmosphere: the volunteers, who feel at home and free to express their requests and their complaints; the children, adolescents and youths, who learn rules and values in a context different from the institutional one (from which the organizations aim to keep their distance, but always with a view to collaboration rather than opposition; and, finally, the adults, often intimidated and embarrassed by a generation of young people who risk being rudderless, without constructing solid biographies in post-modern society—the adults enjoy a more informal environment.

6.3 FBOs at Work: Guidelines from Literature and Empirical Research

In the field of operations, what are the guiding criteria for planning and carrying out the numerous activities of FBOs? Two of them are indispensable.

The first is personalisation. Micro-actions are not defined by an abstract objective but by careful analysis of needs and necessities. In all planning, the intention is to define an educational itinerary conjoining the needs of the territory with the good practices available for dealing with the problem. Thus, every activity may be defined as a training/educational experience of personalisation. However, it must contain four necessary elements:

1. Decisional autonomy (of the operator, in order to organize the intervention in the best possible way);
2. Organizational flexibility (the way in which relations are to be handled);
3. Overcoming an executive mentality (operational tools are prepared by operators and volunteers together since they are the appliers and users);
4. Planning awareness (inserting the activity into an educational path).

The second criterion is feasibility. Unreachable goals are not to be planned for or even less promised. Welcoming, accompanying and promoting acts of advocacy are the result of daily effort, a kind of complex warp, which is woven slowly, incorporating different threads. In doing this there is a risk of adopting positions contrary to local and national laws and normative provisions.

Such criteria orient the arrangement of activities concerning language assistance, school support, and career guidance, and helping to find employment, developing the migratory experience, accompaniment on growth paths and insertion into adult life of young generations as well as aiding the development of the parental role. Thus, in concrete terms, the two principles mentioned above—feasibility and personalisation—help FBOs to orient themselves in the face of the polarity characterizing social work: integrity and functionality, flexibility and rigidity, specialization and generalisation (Table 6.4).

Table 6.4 Overview of the characteristics of pro-social inclusion activities

Integrity	Functionality
Completeness of proposals: from ideation to realisation to verification. Every intervention corresponds with a total educational plan: free time interweaves constructively with school and family time	Educational proposals are shared with other relevant agencies in the territory with which one interacts in a spirit of complementarity
Flexibility	Rigidity
Consistency: a plan of a training/educational offer is backed up with activities; concrete, active proposals with tangible and measurable objectives	There should be a definition of micro-objectives—stages on the path leading to the completion of the educational itinerary—with which to measure oneself in terms of results to be achieved
Specialisation	Generalisation
Consistency of activities with respect to the association's objectives	Activities always have a universal approach

Source: own elaboration Clarke and Ware (2015) and Wejnert and Parigi (2017)

Today, the dimension of the role of FBOs in the management of immigration is significant. This statement is confirmed by the multiplicity of activities, the wide range of active projects and the portfolio of experience gained, the number of volunteers and, above all, the number of children, adolescents, youths and adults who have benefited from the various educational formulae and proposals advanced by the FBOs over time. The following table shows how the activities proposed by religious organisations vary according to the progress of the inclusion process (Table 6.5).

Following Hirschman, the opportunities from which migrants benefit have been grouped into three clusters according to the logic of the 3Rs: Respect, Refuge and Resources: indeed, according to scholars, religious centres and FBOs, they are significant in the migrants' integration paths due to what they can offer in these three domains. It is clear, however, that there is a side effect of the 3Rs: being embedded in ethnic religious environments can be transformed into a segregated and separated arena when migrants develop a parallel life (sometimes full of social recognition) losing interest in investing energies in pursuing social mobility and a full inclusion in the host society (e.g. not improving language knowledge).

In short, it is an important volume of initiatives developed in different sites through interaction and dialectical comparison with varying town and city realities and interlocutors (public and private funders, schools and other third-sector territorial agencies). The challenge of growth crops up. In the field of immigration, it is often a question organizations pose for a specific reason, connected with a particular need, in a well-defined area. Their success and/or new needs to satisfy often impose growth. Growth is a clear indicator of the success of the proposal formula: clear objectives, attention to relationships, and learning through informality. It also testifies to the elasticity of the formulae: their contents are transmitted with adequate modalities and reflect the characteristics of the territory, recipients' needs and the

Table 6.5 Hirschman's 3Rs according to the migrants' inclusion path

Type of need according Hirschman's 3R	Reception	Consolidation	Integration
Emotional-psychological (respect)	Offering opportunities to meet with fellow countrymen, in language Involvement in activities with roles of responsibility	Training activities on parenting in emigration Meetings on the school world	Supporting the care of parents/relatives who have been left behind Maintaining traditions and customs through festivals and celebrations
Informative-legal (refuge)	Support activities for legal integration	Front offices for family reunification practices and recognition of qualifications	Citizenship practices Support for political transnationalism
Material and socio-medical assistance (resource)	Food and clothing distribution Job-search and bed/housing search desk	Places	Care activities for elderly people in their language of origin

Source: own elaboration on Hirschman (2004, 2014)

resources, which can be activated. It is true, as a Barcelona interviewee reminded us that, "*the experience of organizations working with asylum-seekers today is characterized by their managing to anchor themselves in the territory where they developed*". This is reinforced by another trait of the organizations' actions: entering every new situation on tiptoe; which is to say, the ability to perceive and react (rapidly) to external stimuli while uniting two indispensable elements—dialogue and trustworthiness.

The religious-associationism environment is where the difficult task of welcoming, reducing differences in insertion paths, social cohesion and intergenerational support is sometimes carried out. It is one of the pivots of the educational community, which teaches cohabitation and promotes respect and tolerance. In other words, active citizenship is built up within these organizations. This must be done at local level. A further element complicates the situation: the proximity of social transformations. It is in town and city neighbourhoods that the challenge of managing (multicultural and multireligious) change exists. If then it is the task of neighbourhoods—areas of daily reference—and their active educators, to teach (formally and informally) how to decode transforming realities (from the social, instruction and educational paths and employment points of view), it is the task of the citizenry to develop policies supporting those who organize and give life to non-institutional spaces. In this scenario, some of the points of strength and weakness of FBOs may be recalled. A point of strength is succeeding in overcoming the tendency "*to self-referentiality, to closure and isolation, in order to develop interconnection and mutuality abilities*" (Connor 2009, p. 17). Being online and using the web in these years has not just been a simple refrain but a way to relate to other subjects in various urban territories. Such collaboration speaks of capacity for inter-institutional work, from operational flexibility capable of modelling one's own style of work (without distorting it) to real needs very different from one another.

Through their interventions, FBOs succeed in transmitting an idea of possible everyday cohabitation and cohesion. This does not mean that there are never any problems or micro-conflicts, or in unconditional acceptance of difference. Not all operators are on the same wavelength and sometimes differences emerge: who should fund us; whether and how to oppose normative provisions.

Dialectics form part of every organization. There are discussions about whether and how to publicise activities, successes and failures. There is no communications office or even, on a smaller scale, the figure of a fundraiser. In short, a point of weakness is the absence of marketing in terms of managing personnel, members and volunteers as well as public subjects and financing bodies. Then there is the intergenerational question. Who is responsible for it? Who participates in the governance of associations? Both in intercultural and in ethnic organizations, generational rotation causes conflict. It is a power struggle whose solution is found by creating new associations.

The most important lesson emerging from the interviews with FBOs is the importance of a network of associations and institutions working together to define and implement social cohesion in an intercultural way.

As a member of an FBO in Berlin said, "*We don't have a specific intercultural policy: we are intercultural in our way of working, in our activities, in our mentality. We organize the celebration of the end of Ramadan as well as Easter Mass. We discuss the difficulties of growing up as immigrant adolescents as well as discrimination in access to work. The main characteristic of our job is to 'work with', not 'work for'*".

Judging from results, the approach taken has numerous strong points:

- A bi-dimensional approach: initiatives are numerous and various, focussing not only on the reception time span but on the subsequent integration itself;
- Attention to improving the involvement of immigrant/ethnic associations in general projects, not only addressed to single minorities (e.g. Chinese) or specific groups (e.g. Muslims);
- A view at sub-local level: focussing attention on the factual implementation of strategies and not only on their concept/promotion, allowing the municipality to understand better the difficulties of immigrant associations and administrators when defining strategies, interventions and policies within the societal area;
- The definition of interdepartmental meetings and projects: a comparison with workers coming from different fields enables the municipality to better understand the potentialities and shortcomings of the strategies implemented in various societal fields;
- The involvement of NGOs working with immigrants allows the municipality to investigate better the needs and problems of foreign people and to evaluate the proposed strategies.

It is important to continue to support programmes concentrating on the following core areas: language acquisition, education, integration into the labour market, social counselling and social integration. Nonetheless, the integration programme is not a one-time project but a continuous, dynamic process open to address other issues as they become important. In this way, projects intended to promote civic education, participation and solidarity among the second generation represent a willingness to think beyond current problems and aim towards the future.

6.4 The "Refugee Crisis" and the Third Sector: The Italian Experience as an Explanatory Case

The Mediterranean context is characterized by a great variety of subjects that deal with the migration issue: among these NGOs and associations, both ethnic and religious. The tradition of supporting people in trouble is rooted in the experience of several religious orders and FBOs. This experience helped organizations to accept the challenge that the new immigration started to present. This was a challenge that religious orders and, primarily, Catholic volunteer organizations tried to face, sometimes spontaneously, at other times trying to adjust to the needs of different cultures, modifying their methods of intervention. Since the mid-1980s, many ethnic associations have come into being. The role of FBOs is becoming crucial in managing migrants' issues. Again, Italy appears as an interesting case study, whose findings could be generalized in a Southern European model. Indeed, various town and city departments have involved associations in improving their services and activities. This methodology is defined as an "un-institutionalized method": spontaneous networks have developed in order to deal with specific emergencies: asylum-seekers from the former Yugoslavia, unaccompanied minors, and trafficking of women. Each department has a specific association network with which to collaborate in implementing activities and promoting projects.

The management of migratory phenomena by local authorities could be synthesized in three main periods:

1. The emergency policy (1990);
2. The creation of a network of specific services, based on the interaction between the municipality and the third sector (1990–1995), implementing an interesting "welfare mix" model;
3. A period of consolidation since 1995. This phase can be divided into two subperiods: (a) 1995–1999, during which municipalities improved their relations with associations, supported the insertion of cultural mediators in their offices, developed training courses for civil servants, and expanded their attention to the migration issue, from labour to family services, from school to welfare support; (b) 2000 to the present day. In this time-span, there has been a consolidation of activities in the "most intercultural areas" in order to reinforce positive relations between autochthonous people and immigrants. Attention has also turned to the integration process of the second generation, accompanied by a necessary focus on the increasing number of asylum seekers.

In general, this "ad hoc approach" has meant that municipal interventions were explicitly directed towards immigrants in specific difficult situations, reacting to emerging social problems by mobilizing all the main local actors, public and private, interested in the specific issue.

During the last 20 years, however, the consolidation of intercultural discourse has taken place in tandem, generating a large number of practices, projects and experiences based mainly on the notion of dialogue, mutual exchange and social interaction. Intercultural policies are considered as part of the general integration strategy for immigrants. In this way, the creation of Intercultural Centres is significant regarding the strategic role assigned by the towns and cities to initiatives aimed at promoting awareness among all citizens of changes and new challenges resulting from the evolving composition of the resident population. The idea is that each policy should be defined while taking into account the various heterogeneous intercultural facets existing in the towns and cities.

By involving associations, NGOs and immigrant communities, local authorities have developed projects, initiatives and policies. In addition, in recent years, these projects have been managed together with the asylum crisis.

In Italian towns and cities, at local level, during the refugee crisis the influx of asylum seekers and refugees from countries with Muslim majorities offered opportunities for shared activities between Muslim and Christian Catholic organizations, strengthening the Muslim and the European/Italian identity of both Muslim refugees and Muslim refugee workers (Khallouck 2018). In fact, Islamic associations[2] too in Southern Europe run programmes for refugees at various levels, covering all areas of refugee relief such as first aid, beds, food and guidance and support activities—for example, during "*Emergenza Siria*" (asylum seekers mainly from Syria aiming to go to Germany but stuck at the Central Station in Milan) in close coordination with other national, local and religious (mainly Christian Catholic) associations. The contexts in which some of the refugee workers were refugees themselves can also help them to overcome difficulties through proactive participation in public and political life, following the role model of those who "made it". The shared history of fleeing creates trust between the refugee workers and the refugees who are considered equals.

Often, for the refugees and asylum seekers, being in a new reality, without familiar linguistic-cultural references, can favour seeking refuge in religion, perceived as an element of recognition and reinforcement of identity, but also as a cure for loneliness and isolation. This is mainly because attendance at churches, mosques and places of worship allows meeting other migrants from the same country with whom one can share experiences and needs (Jackson and Passarelli 2016).

The worship centres (churches, prayer halls, mosques, religious associations), especially during the so-called refugee crisis, were confirmed to be a point of reference, carrying out those functions typical of religious organizations in emigration, not only as a religious reference, but also (and especially) as an identity reference (Levitt 2007).

> I feel well when I am in church. I think of my sisters and brothers at home and get a little homesick. We all often went to church together. Here, I'm the only Catholic in the welcoming center, all the others are Muslims or Pentecostals. At the beginning of my period in Italy, I thought I had to be grateful to be in the Pope's country. Sunday-by-Sunday I discovered to what extent Italians are different in their being Catholics from me. I felt alone during Mass, as a person coming from another planet: religious rites here are without joy and participation. People seem to attend Mass just to tick another task in their diaries. Now, I used to attend the African-community Mass and I felt at home. Of course, all the churchgoers are foreigners, Africans (F, Nigerian, representative of an intercultural association, Rome).

Again, at local level, some inter-religious experiences arise. The most common example deals with the organization of *iftars* (fast-breaking) during Ramadan or shared/open to the public celebrations inviting local and regional politicians, representatives of other religions and refugees in order to bring together Muslims and non-Muslims (Premazzi 2017; Khallouck 2018). These activities allow mainly vulnerable migrants (e.g. asylum seekers) to feel welcomed and concretely experience a sense of community in their new host society. In these shared practices, "they

---

[2] Islamic Relief has put forward *'a three-point Agenda for Action'* to tackle the crisis, including a Europe-wide humanitarian response as refugees travel across the continent, greater commitment to resettling refugees in European and other countries, and a fresh diplomatic effort to end the conflict in Syria, providing food packs to refugees, as well as small cash grants to the most vulnerable, and translation and support services to help people get the assistance they need.

encounter a nuanced experience of Italy that, on the one hand, prevents naivety, and, on the other, anxiety, with reference to society (Premazzi & Ricucci, forthcoming).

> We have never worked with so many associations as we do now. What happened has also had a positive aspect in that it drove us to meet, to talk to, to understand one another. Its weight is most certainly felt in the media; for example, a lot of nonsense is said on talk shows, creating this climate where people come up to you and ask: 'And what do you think about ISIS?' And 'what do you think about terrorists coming from Libya?' And 'what do you think about refugees', yes, that happens, but on a personal level we have not observed any substantial change of attitude. Rather, there has been even more coordination with the security forces. Perhaps the only positive thing all this has given us is the fact that it has driven us to converse, to collaborate and to get to know one another better. There has never been such collaboration and dialogue among different religious organizations and people before. There is also talk about a conference of religions here in Turin, different projects, which are being worked out (Islamic representative, Turin).

6.5 Religious Experience Far from Home: From Temporary Guests to Civic Actors

In an age marked by serious problems with foreigners' otherness, we cannot claim that processes of cohabitation and safeguards for the diversity of religious minorities are created spontaneously, without paying particular attention to social-cohesion processes and accompanying the protagonists—which is to say (recalling Hammar's classification), everybody: citizens, semi-citizens, immigrants and foreigners (1990). First among these are the young generations, destined to be the focus of planning, interventions and policies. New needs come under the spotlight and specific groups interrogate public decision-makers and civil society about possible answers—solutions that demand redefinition of the borders of belonging, of recognition and even of national belonging (Foner and Simon 2015).

Foreigners, and even more, their children, update a difficult task, which began in the distant past when the movement of peoples constructed a European panorama of which many are today seeking a different, updated, version. Young Muslims and their associations seem to be the most visible novelty—not the only one, however, nor the most numerous. Indeed, it is well to remind ourselves once again of the gradual Europeanization of flows and the effects of the economic-financial crisis, which have re-drawn the map of diversity in Mediterranean countries, and to be on our guard against excessive (and fallacious) simplification (Triandafyllidou and Ulasiuk 2014).

How can one insert generational transformations on the religious front into the dialectics between demands for recognition and local institutions? What is the attitude of great and small towns and cities to a generation of young people of foreign origin, with variegated religious affiliations, appearing on the public stage aware that they have rights in that they are residents and citizens?

The time has now passed when young people marched in the streets and carried out emblematic acts. In the age of social media, activism in the religious sphere too takes place on the web: it is propagated and strengthened through websites and social networks. However, associationism, with its capacity to promote and represent common demands, remains central.

Awareness, knowledge and involvement are the three key words synthesizing the actions of Muslim associations in the European Diaspora countries: awareness of being under observation by at least three groups: that of local natives; that of their origins; and that of the transnational Ummah.

Attention is dedicated to each of these by means of events using their own language and themes. As Ricucci and Premazzi (2017) revealed, in this task, digital skills and the opportunity to construct transnational networks meant a card up one's sleeve compared with the previous generation's ability to act.

Self-knowledge is another important element. Young people know they have to confront events transcending their daily lives. They know that every initiative where Muslims (including second generations) take the lead casts a shadow over their work, an additional obstacle to be overcome. The knowledge that being "*on the right side, with the language and tools of every honest citizen, is our strength*", to quote a young Muslim during a demonstration in support of Islam and mosques in the city of Milan.

Finally, involvement. From Rome to Madrid, Brussels to Lisbon, many young people involved in Muslim associationism define themselves as active citizens. They volunteer for socially useful activities, and take part in intercultural events, religious dialogue and social advancement originated by administrations, private bodies and—religious or lay—commitment groups (Demichelis 2018).

Involvement in the place where one lives is part of a true strategy for building credibility and trust that they are, at the same time, committed and believers, and, in some cases, young people committed *qua* believers (Table 3.2). Typology of relations between associationism and local institutions

Like all models, the one above crystallizes positions, which are modified over time by the effects of variables which may be contextual (policies) or depend on the people involved—*Who holds the power in associations? What generation(s)?* However, arriving only occasionally at direct involvement in policy-making processes (De Graauw and Vermeulen 2016). In most Mediterranean contexts the dynamics of involvement and partnership in the sphere of intercultural activities, and those of a consultative type, seem to prevail.

For years, political decision makers have paid attention to recognizing the children of immigration as key actors in building social cohesion and dialogue between natives and immigrants. In this way, young foreigners (or of foreign origin) are supported in their activities as involved citizens, in the organization and promotion of autonomous associationism activities.

They are considered important actors in building or strengthening social-cohesion processes, especially in the areas where native-immigrant interactions are most difficult and where polemics about cultural and religious diversity flare up again year after year fuelled by higher visibility, by a process of gradual gaining of citizenship, by the increasing number of students in schools, and by demands of Muslim associations. This situation is interpreted not only as an opportunity to intervene but also to support associative leadership with a view to encouraging the idea of "active citizenship" in an environment where there is also room for the religious dimension.

A diachronic view allows us to emphasize points of strength and weakness in first- and second-generation associationism in their relations with local administrations. The first generations had a primarily on-demand relationship: it was necessary to reconstruct prayer infrastructure, find spaces, and obtain authorization for the ritual slaughter of animals and community celebrations. When migratory seniority increased, suitable, recognized burial spaces also became necessary. Local administrations, for their part, focussed on activities encouraging tolerance and respect for human rights, awareness of the positive value of diversity, increasing intercultural contacts and the development of ethnic and interethnic associations.

Relations with the youngest shift towards partnership, in the latent hope that for the children, themes such as ethnic-religious-based identity claims, and requests for exceptionality due to beliefs and practices rooted in religion, will decrease and lose some of their symbolic power.

New generations aim to be recognized as partners, even in competition with their parents' generation. Their objective is to be present and active in cultural events, bearing witness and competent voices—not the often stereotyped, functional vision of socio-cultural environments where religion (and, first and foremost, Islam) is the main point of reference. Whenever possible, they ask questions and intervene in decision-making processes in order to reinforce the thesis that a religious message is compatible with good citizenship. Even in towns and cities with the most advanced integration policies at local level (Caponio et al. 2018), these aspirations clash with reality: second-generation associations—albeit appreciated by many local political groups—are not yet ready to be perceived as being capable of substituting their parents' associationism as a point of reference for institutions. However, we should not confuse absence from the political, decision-making process as total absence from the debate concerning these questions of interest: the voice through which the children of immigration intend to define religious affiliation will be broadcast by means of attendance at congresses, organization of public events and, above all, the web.

References

Adogame, A. (2007). Religion in Sub-Saharan Africa. In P. Beyer & L. Beaman (Eds.), *Religion, globalization and culture* (pp. 533–554). Brill.

Ambrosini, M. (2015). NGOs and health services for irregular immigrants in Italy: When the protection of human rights challenges the laws. *Journal of Immigrant and Refugee Studies, 13*(2), 116–134.

Caponio, T., Ponzo, I., & Ricucci, R. (2018). Policy change or policy re-styling? The case of intercultural integration policy in Turin, Lisbon and Valencia. In M. Sacco, C. Torrekens, I. Adam, & F. Zibouh (Eds.), *Circulation des idées et des modèles: les transformations de l'action publique en question. Le cas des politiques d'intégration* (pp. 123–137). Academia-Bruylant.

Carella, M., Gurrieri, A. R., & Lorizio, M. (2007). The role of non-profit organisations in migration policies: Spain and Italy compared. *The Journal of Socio-Economics, 36*(6), 914–931.

Carrette, J. & Miall, H. (2012). *Big society or global village? Religious NGOs, Civil Society and the UN*. Briefing Paper.

Castelli, J., & McCarty, J. (1997). *Religion-sponsored social service providers: The not-so-independent sector*. Aspen Institute Nonprofit Sector Research Fund.

Clarke, M., & Ware, V. A. (2015). Understanding faith-based organizations: How FBOs are contrasted with NGOs in international development literature. *Progress in Development Studies, 15*(1), 37–48.

Connor, P. (2009). International migration and religious participation: The mediating impact of individual and contextual effects. *Sociology Forum, 24*, 779–803.

Crawley, H., & Skleparis, D. (2018). Refugees, migrants, neither, both: Categorical fetishism and the politics of bounding in Europe's "migration crisis". *Journal of Ethnic and Migration Studies, 44*(1), 48–64.

De Graauw, E., & Vermeulen, F. (2016). Cities and the politics of immigrant integration: A comparison of Berlin, Amsterdam, New York City, and San Francisco. *Journal of Ethnic and Migration Studies, 42*(6), 989–1012.

Demichelis, M. (Ed.). (2018). *Narratives of Islam Christian and gender violence in contemporary age*. Olms.

Foner, N., & Simon, P. (2015). *Fear, anxiety, and national identity: Immigration and belonging in North America and Western Europe*. Sage.

Garelli, F. (2020). *Gente di poca fede*. Il Mulino.

Georgi, F. (2019). The role of racism in the European "Migration crisis": A historical materialist perspective. In V. Satgar (Ed.), *Racism after apartheid: Challenges for Marxism and anti-racism* (pp. 96–117). Wits University Press.

Georgiana, I., & Agafiţei, M. (2016). First and second-generation immigrants - statistics on labour market indicators. *Eurostat*. Retrieved 20 August, 2020, from http://ec.europa.eu/eurostat/statistics-explained/index.php/First_and_second-generation_immigrants_-_statistics_on_labour_market_indicators

Hirschman, C. (2004). The role of religion in the origins and adaptation of immigrant groups in the United States. *International Migration Review, 38*(3), 1206–1233.

Hirschman, C. (2014). Immigration to the United States: Recent trends and future prospects. *Malaysian Journal of Economic Studies: Journal of the Malaysian Economic Association and the Faculty of Economics and Administration, University of Malaya, 51*(1), 69–85.

Jackson, D., & Passarelli, A. (2016). *Mapping migration, mapping churches' responses in Europe*. In *Belonging, community and integration: The witness and service of churches in Europe*. World Council of Churches and Churches' Commission for Migrants in Europe.

Jeandesboz, J., & Pallister-Wilkins, P. (2016). Crisis, routine, consolidation: The politics of the Mediterranean migration crisis. *Mediterranean Politics, 21*(2), 316–320.

Khallouck, M. (2018). Confronting the current refugee crisis: The importance of Islamic citizens' initiatives in Germany. In U. Schmiedel & G. Smith (Eds.), *Religion in the European refugee crisis* (pp. 87–103). Palgrave MacMillan.

Levitt, P. (2007). *God needs no passport: Immigrants and the changing American religious landscape*. The New Press.

Longo, F. (2019). Italian migration policy between perceptions and practices. *Contemporary Italian Politics*. https://doi.org/10.1080/23248823.2019.1677402

Premazzi, V. (2017). *Internet, Islam and second generations: Risks and opportunities*. A-id Agenda for International Development.

Tadros, V. (2017). The persistence of the right of return. *Politics, Philosophy & Economy, 16*(4), 375–399.

Triandafyllidou, A., & Ulasiuk, I. (2014). *Diversity partnerships. Towards a common framework for migrants and minorities*. Green Paper, Global Governance Programme, Policy Workshop.

Vidal, A. C. (2001). *Faith based organizations in community development*. HUD's Office of Policy Development and Research (PD&R). Retrieved 1 December, 2020, from https://www.huduser.gov/portal/publications/faithbased.pdf

Wejnert, B., & Parigi, P. (Eds.). (2017). *On the cross road of polity, political elites, and mobilization*. Emerald Group.

Wuthnow, R. (2005). *America and the challenges of religious diversity*. Princeton University Press.

Chapter 7
It is Time to Change Perspective: Outside the Framework—Muslim Exceptions

In the post-2001 climate, religion returned predominantly on the agenda of governments, even those bulwarks of the concept of secularism. Today, Europe stands out from the US for its inability to manage the particular diversities of religions and their interpretations. Unlike the past, it is no longer a question of internal divisions within Christianity, but of relations with the variegated world of faiths and their declinations.

Emblematically, Bowen, in 2009, wondered if a French Islam was possible, focussing on the ambivalence between an everyday life steeped in religious references in which Maghrebis, Senegalese, Ivorians, and other French citizens of foreign origin who define themselves Muslims, and a dominant narrative of secularism and secularization. Above all, not exclusively in the French scenario, a dystonia emerges between the script recited in public (the absence or irrelevance of religion in public spaces or even in the life of individuals) and the real canvas, that is, everyday life imbued with references of faith, from architecture to music, from shops to clothing and to the manifestations of key moments of the numerous religious calendars. We also ask ourselves how secular Europe deals with issues related to the multiple forms of faith, in its renewed leadership.

Starting from these interesting and legitimate questions, there is much debate about religious pluralism. Yet, with careful attention, one immediately understands how only some confessions represent real objects of discussion. It is, above all, Muslims who end up under observation at every European latitude, as described in Chap. 4; more than in the United States, according to the most recent surveys (Evans and Baronavski 2018). On the other hand, already in 2006 the counter-intuitive result of a survey by the Economist magazine found that Muslims (and members of other religious affiliations) felt more protected in the USA than in Europe, in particular in the United Kingdom, an emblematic country of multiculturalism on the Old Continent.

In short, the main differences appear for Americans to be traced back to a history dotted with examples of confrontation with different religious groups, a citizenship that continues to consider itself religious in its majority, and the commitment of

R. Ricucci, *Ethnicity, Identity and Faith in the Current Migratory Crisis*,
https://doi.org/10.1007/978-3-030-84056-3_7

advocacy groups of religious minorities to make known traditions and values, in tune with what happened in the past and is in the DNA of American history. A fundamental part of this scenario is the relationship—today more complex and difficult to manage—with Muslims. Thomas Jefferson wrote that the key test for America's success as a land of freedom and equal dignity for all would be welcoming the followers of Muhammad, at the time considered truly "*the other*". However, these positions are not sufficient to explain the different attitudes towards the faithful of the Prophet on the two sides of the Atlantic. Migration profiles, selective reasons for entry, geographical distance and different ways of entering the community of citizens can help to reconstruct the reasons for a more positive attitude towards Muslims in the US. We should not forget the serious acts of violence that have occurred over time (such as the recent massacres in Orlando and San Bernardino), as well as those arrested for praising ISIS, or the presence of mosques where outbreaks of extremism lurk. In general, however, data show the an entry of the majority of Muslims into mainstream American society. These Muslims agree plural interpretations of the way of living Islam are possible.

In Europe, relations with Muslim communities are decidedly more complex and varied, country by country, local context by local context. In fact, the discussion on diversity emblematically becomes a debate on immigrant citizens in Europe, thus charged with prejudices of an ideological type. More, and in a more circumscribed way, it becomes a discussion on Islam. Even when the faithful do not represent the majority of immigrants nor are they only attributable to foreign citizens, but are part of European citizenship, with a passport from one of the member states of the European Union. In fact, by recalling what Pace (2004) said, Islam in Europe "*re-knots the threads of a memory broken by history. This means that the Muslim presence will be one more reason to reflect, not only on the identity of these European Muslims, but also on the historical roots of the identity of Europe itself and on the fractures that have divided the Mediterranean basin for centuries*" (p. 20).

How are memberships in non-European national communities connected with the duties of every Muslim of belonging to the *ummah*, or rather, to the Islamic supranational community? What are the possible irreconcilable issues between *shariah* law and legal, economic and social systems consolidated in Europe and supported by standards, including international ones? More than 30 years after the first dispute over the veil worn at school in France, numerous episodes have challenged the constitutional framework of the major immigration countries, highlighting the fragility of the structures that govern the relationship between churches and public institutions. The religious calendar, school teaching, the construction of places of worship, the recognition of ministers of worship in public places, as well as hospitals, barracks, prisons; the possibility of wearing specific clothing, prayer-times, legal recognition of organizations and their funding: these are the issues that question a coexistence with precarious foundations. Here, the Muslim performs a mirror function, showing the aporia of compromises and the anachronism of institutional architectures challenged by human mobility and the progressive stabilization and structural presence of immigrants who—by conviction or cultural adherence—have a point of reference in Islam for religious, economic and social life.

One final line should be dedicated, due to its relevance and modernity, to the discussion of the right to religious freedom. In Europe, a common model of this right has been taking shape "*mainly focussed on the individual (freedom of conscience and religion) but with the recognition of a collective dimension, also institutionally organized, as confessional autonomy*" (Angelucci 2017, p. 54).

While respecting human rights and fundamental freedoms, even in the religious field the outcome is a Europe of nations, where the articulated systems of relations between the state and churches maintain their strength, only partially conditioned by the rulings of the European Court of Human Rights in Strasbourg, whose work is "*actually becoming the point of reference in matters of fundamental rights for the same organs of the Union*" (Mazzola 2014, p. 1). The result is in fact a heterogeneous situation, with experiences of acceptance or latent discrimination, protection or case-by-case evaluation, found in the various European countries (Soper and Fetzer 2005). In the dialectic (not always devoid of conflicting elements), between administrative authorities and religious organizations, five variables emerge as relevant and interconnected (Fig. 7.1).

However, the factors that come into play in the management of religious pluralism are not disconnected from social scenarios, historical-juridical traditions and the political forces involved. These last three variables affect how the national legislator—in harmony with the recognition of international norms—recognizes religious rights to actors and minority groups. The process is neither smooth nor free of tension. Lobbying activities (partly through governments and international pressure groups) confront each other in the political and, above all, social arena to protect interests or prevent opportunities for recognition from increasing (Berger 2015). Religions thus become the mirror of the ability of societies to confront and manage diversity.

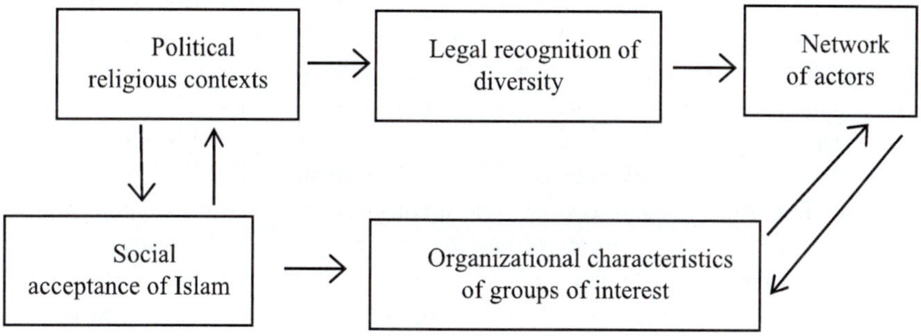

Source: Peña-

Fig. 7.1 Variables of the relationship between administrative authorities and religious organizations. Source: Penninx et al. (2004)

7.1 Implications for Research and Policy

The growth of a second generation and the number of those who become German, Spanish or Italian are two elements that mark a symbolic change and concrete effects within the European scene. A transformation that concerns the direct protagonists (the young) and reverberates in their relationship with society, translates into the relationship with the congregations that host the ethnic chaplaincies, in the debate on the need for ethnic churches, in the confrontation with the many identities that make up the Catholic, Orthodox and Evangelical world. The entry into adult life and into the community of citizens of the children of immigration invites scholars and professionals, service providers, teachers and residents to become aware of the fluidity and the many possible variations in terms of religious identities. In this sense—and also on the side of faith—the interviewees give voice to a reality that no longer wants to be in the spotlight because it is foreign or traceable to practices and rituals of other geographical contexts, but to be considered real to the generation of young people who now study at university, seek work, try to become adults and build an autonomy of living and income. In short, even in religious matters, the desire to be "*treated like Europeans and as Europeans*", the desire to "*believe or not, attend weekly or rare functions*", the request to "*stop being children of foreigners, children of immigrants, children of Moroccans or Senegalese. We are boys and girls who live here and we ask to be judged for what we have studied, for what we do, not for what we have inherited, including religion*".

In addition to the changes in the way of positioning oneself faced with the themes of sacredness, in the towns and cities there is also another transformation. The number of ethnic churches and places of worship has increased, as well as the number of religious associations and activities promoted and carried out directly by ethnic-national cultural centres or in partnership with other local realities, institutions or subjects of civil society. In fact, in recent years the calendar of proposals for meetings, debates, interaction with the entire citizenry with a view to openness and knowledge, has been enriched, with some locations more active than others. A cultural liveliness that denotes, on the one hand, a desire for participation and protagonism within the context of one's own life; on the other, there is a positive tension to come out of one's surroundings to establish relationships, and relationships with other religious groups living in the same environment or to participate in initiatives of solidarity or an inter-religious matrix (Martínez-Ariño 2017).

In addition to the Orthodox popes and imams trained in Europe in order to give an answer to the request of control and the fear of importing dangerous religious speakers, a new element in the social context of many European towns and cities, particularly in Southern Europe, there has been an increase in the number of priests from the countries of origin of the most important Catholic communities, and their visibility and presence within the dioceses has increased, still often concentrated in the field of pastoral care for migrants. A novelty, which offers the opportunity to carry on the religious demands of minority groups and yet reaffirms how this pastoral care continues to have to maintain special features, which makes it different from that of the rest of the community. It is certainly too early to grasp the signs of religious integration, with the widespread participation of Catholics of foreign origin in pastoral councils or in the numerous activities of parishes. The transition from the cohabitation of the church space to a real sharing and community organization has yet to take place. It is not an easy path. For the "native" believers (Spanish, Portuguese, Italian), who fear they will have to integrate their habits, ways of managing sacredness with a religiosity that seems to most people popular, typical of other times, and not in keeping with the rhythms and customs of modernity. For immigrants, who do not want to abandon their language in the liturgy, to lose a privileged relationship with a priest who—in many cases—lived in their country, and knows the reality and can well understand the concerns of their parishioners. In the middle is the theme of the young, of the second generations, who risk becoming hostages of one or the other. In one case, attracted by the activities of parishes (either mosques or temples), where participation has little to do with religiousness and much to do with the tension to assimilation, which aims to "not feel different"; in the other, linked to the family and its religious socialization, which transcends the boundaries of sacredness to become also a socialization to the values, language, and traditions of the country of parents.

The youth component represents a possible factor for innovation within FBOs or ethnic religious associations. In particular, five factors appear to be crucial to making these activities fit into the landscape of each local context, responding to the expectations of policymakers, who believe that "*faith-based organizations might take on a greater role, offering multiple sources: the desire in some quarters to reduce the role of the public sector; the existence of a small number of high-profile successes in housing and economic development sponsored by large churches (or mosques…); the perceived paucity of other strong institutions in many disinvested neighbourhoods; and high expectations about the potential of faith communities to address problems that others have found intractable*" (Vidal 2001, p. 1).

In order to achieve these goals, young people could be an important collaborative factor. Indeed, the juvenile perspective can understand better the current needs for improving social cohesion than the adult one, mainly focused on the development of the country of origin, and replace in emigration what has been experienced elsewhere. It could be difficult to conquer space and to raise one's voice in first-generation environments, where youth are welcome for their numbers, but their ideas are ignored, as being too revolutionary, as it happens in each game among generations, independently of migration matters. However, young people can challenge these organizations, helping them on the following issues:

1. Ethnic compositions: it is time to overcome the ethnic homogeneity which could be necessary in the first phase of welcoming. If inclusion is the goal, *mixité* and interaction is the way to avoid the risk of ethnic segregation;
2. Develop strong internal management and governance: it is time to improve the governance of these key organizations, whose help is crucial and recognized by local actors. Due to this, a good management structure is needed and, on this point, young people with their language skills could be helpful (Ricucci 2021);

3. Secure resources: fund-raising activities are crucial in each organization, and the skills in acquiring grants, loans, fees or earnings, and technical assistance should be present in each VolAg (Voluntary Agency), such as in the FBOs and ethic religious associations. Young people could insert new ideas and develop initiatives thanks to their juvenile perspectives and their global views, better interacting with their co-peers and understanding the mood of the numerous public and private bodies addressing funds to the youth component within migrant communities;
4. Network: the skill to interact with all the local actors as well as to develop ties transnationally is becoming important. In the digital era, young people are at the forefront.

Considering the above-mentioned requisites, the role of the second generation appears to be a keystone. They are "*sought-after and welcome*" because they represent the guarantee of numbers and integration into society, for their ability to interact with local administrators, supranational organizations and participate in European events (Ferrari and Pastorelli 2012). Nevertheless, for all religious affiliations, the second-generation associationism linked to faiths begins to become the protagonist of a search for relations with the social and cultural context in which it operates. For example, information initiatives are carried out for citizens in the community and its value system, information materials are produced on the main festivals and events, and the meaning of religious rites, and initiatives are taken to meet ecumenically or interreligiously.

7.2 Cities, Crossroads of Migrations and Religious Affiliations

Taking up what Kivisto (2014) wrote about the Brick Lane area in London, several city districts also presented themselves as areas that had experienced different migratory cycles (in Italy as well as in other countries) and are now vivid multicultural and multiethnic centres (OECD 2018). At the same time, they are areas where first and second generations began to live together, where experiments with schools on inter-religious education and community welfare activities promoted in collaboration between churches and musallas began; in other words, microcosms in which work was not easy among the different actors of the territories. Indeed, all the districts seem to have followed the same path, as the Table 7.1 shows.

In the 1990s, the lack of social networks among people sharing the same neighbourhood was dangerous in managing social and cultural conflicts and to avoid the issue becoming exclusively one of security. In urban areas, the municipality adopted a holistic approach to improve the quality of life for everybody (e.g. housing, economy, social life, cultural initiatives, renewing public spaces, and new functions for buildings), adopting a "*community development approach*". The main goal was to build a social compact in order to promote a new social cohesion process based on

Table 7.1 Issues posed by ethnic religious associations to local administrators. Comparison among four Southern European cities

	Athens	Lisbon	Turin	Barcelona	Montpellier
Recognition as part of the decision-making process		X	X		
Funds for associations' activities	X		X		
Space for meetings	X	X	X	X	X
Religious buildings	X		X	X	X
Specific areas for burials	X	X	X	X	X
Religious education at school		X	X	X	
Halal food in the canteen in public services (schools, hospitals, jails)		X	X	X	X
Promoting initiatives against religious discrimination	X	X	X	X	X
Promoting/organizing Arabic language teaching at school	X			X	X
Public space for religious activities (e.g. for the celebration of religious holidays)	X	X	X	X	

Source: Own elaboration based on interviews with stakeholders and fieldwork in these cities in the period 2016–2019

collaboration among the various formal and informal organizations living in the area. These included parishes, ethnic associations, local resident committees, entrepreneurship committees, and youth and elderly groups. In the process, different town and city departments were involved in partnership with actors and organizations from the area of civil society and the non-profit sector (OECD 2018; CLIP 2010). This policy was set up in collaboration with various associations (several of them FBOs) and NGOs working on the issues of urban policies and social cohesion. If they were to play an active role in the transformation of neighbourhoods, the empowerment of people, inhabitants and local actors was still the main task of the local agencies.

Yet, despite the practices and efforts, suburbs and districts known as "the multi-ethnic areas" continue to be in the imagination of many the symbols of those who cannot be integrated into the local, European societies, primarily into the religious arenas. Changing such imaginaries is not easy: numerous programmes in the countries of the oldest immigration have been promoting this for over 30 years. Various experiences, from those in the French suburbs to those in Sweden, have shown that only by involving the various dimensions of integration policies (education, labour, welfare, health, housing, urban planning, economic development and so forth) can results be achieved. Yet, if it is at local level that policies and partnerships can be built to promote integration, it is also at local level that the tension between those who consider immigrants as the *pars construens* of European society and those who consider them as the *pars destruens* (competitors for labour resources, subsidies and access to rights) can be seen. This dialectic emerges above all in realities where the presence of foreign citizens, in absolute values and as an incidence on the total number of residents, is most significant. Moreover, they are often realities where there are difficulties of social inclusion in the suburbs, which existed before the arrival of immigrants. The cities, from Turin to Valencia, Milan to Barcelona, Rome to Athens, and Genoa to Lisbon, are multicultural and multi-religious areas. They are also areas where the logic that guides many European citizens is best understood: migrants are "wanted and rejected". In other words, they are accepted if they are engaged in public service activities or work in the countryside (this happens, for example, in seasonal work), but removed from view, far from the neighbourhoods, from the eyes of women and children. It is a logic that is transversally applied in all the local contexts of southern Europe, which are now confronted with their multiculturalism, to the requests of Muslim associations to build places of prayer. The relationship with Islam represents a challenge; one could almost say "*the* challenge"; not so much because of the demands but because of the entry onto the public scene of the children of the first pioneer Muslims from Morocco, Tunisia, Egypt and Senegal. The internal dialectic within Muslim associations between the first and second generation is reflected externally in the search by the younger people and more often oriented to develop a European Islam, a way of living the religion that takes into account the context and their socialisation, often all lived in the countries of the southern border of the EU.

Often memory wavers; polemics are kindled and invigorated following the entangling of arguments and rhetorical messages that poison reality and undermine laboriously constructed processes of social cohesion. In fact, the debates about the relationship between religion and second generations that have attracted the attention of scholars since the attacks on the subways of London and Madrid, up to the most recent tragic events in France and Belgium, seem already far away. Yet the elements of the question seem to be the same: youth, immigration, integration, religion.

The international conflicts, ISIS, the explosions in the heart of Europe and the elements of discrimination and prejudice that Muslims know today in their numerous diasporas place the relationship between the children of immigration and the religion of Mohammed in the spotlight. Nevertheless, the second generations are not just those linked to Islam. Especially, in the Italian scenario, where the presence of Muslim immigrants is surpassed by the Christian one (UNHCR 2019).

The polycentrism of migration, characteristic of the Italian context, i.e. the coexistence of a significant plurality of communities and not only of one or two strong national groups (as in the case of Germany with the Turkish presence), is also found with regard to the religious landscape. The activities for young people organized by the Sikhs in the temple of Novellara, one of the largest in Europe, or the Children's Summer promoted by Islamic prayer rooms, which is entirely comparable to that organized by Filipino or South American chaplaincies, are examples that introduce the debate between religion and integration, which is increasingly alive on the individual territories, in the arena that involves citizens, administrations and immigrant communities. The game of inclusion is played on the field of integration policies at local level also with regard to the relationship with faith: for adults and young people, first and second generations. Towns, cities and neighbourhoods obviously not only differ in the composition of the immigrant population, socio-demographic characteristics, initiatives and opportunities offered, but also in the way they relate to citizenship in general and the resulting processes of social cohesion. Among these strategies are also those related to the management of religious diversity, which often ends up being reduced to relations between the towns and cities and the Muslim presence (Allievi 2010). Again, the Italian case could be used as an explanatory case of the extent to which relations with Muslims are still far away from being conflict-free among local residents (neither Muslim nor immigrant) and those who identify with Islam.

7.3 Patterns of Perception of Conflictual Lines of Religious Differentiation

We are accustomed to study Italian perceptions of Muslims living in the country using data from a survey conducted in 2019 on the religiosity of Italians. This research detected the opinions and attitudes of Italians towards certain practices generally considered problematic for public opinion. As shown by numerous studies in Italian, the words "foreign" and "Islamists" have become in the last decade almost synonymous. According to the survey, 86% of Italians are Catholic, 11% with no religious affiliation, and 3% belong to a religion other than Catholic. In total, according to this survey, Muslims slightly exceed 1.5% of the total population, being therefore statistically irrelevant. In Italy, the non-Catholic religious communities are numerically insignificant; the situation is changing, but talking about the Italian attitude towards Islam means in most cases speaking of the opinion of Catholics or those without any religion.

Only 23.1% of Italians had a cultural interest (historical and socio-economic aspects) in Islam and this percentage dropped to 3.5% with respect to the spiritual interest. These data do not differ much from those observed for other non-Christian religions (e.g. Buddhism, Judaism or New Age) but it indicates a lack of interest among Italians in other religions.

However, many other questions arise in trying to understand the perception of Islam among Italians. Among many topics, three in particular can represent a litmus test of Italians' opinions about possible cohabitation with Islam: the education question, the rights of women, and the interweaving of religious practice and the timetable of public life.

The question of education relates to the request—forwarded by the Islamic community—that Muslims living in Italy should be educated more in line with their history and culture than the other Italians. The population as a whole continues to have a negative attitude towards this need: 43% of the population are strongly opposed to opening Islamic schools or ad hoc education for Muslims, another 43% are not sure and 14% are in favour. The relatively greater open-mindedness of Italians in this field seems to be a recognition of a greatly evolving reality. On the one hand, we have data about the increasing birth rate among immigrants and greater numbers of non-Italian pupils, calling attention to a rapidly growing second generation, which needs to be educated. On the other, episodes in Milan of exclusively Islamic classes in a social-sciences secondary school and the opening in 2006 of the Italo-Egyptian "Nagib Mahfuz" institute (an Egyptian consular school of Islamic culture) remind us that education is a field of continual comparison/contrast not only in public debate but between autochthonous and allochthonous people. The dispute regards both the degree of "mixedness" (the proportion between Italian and immigrant origin students), which is considered acceptable in an Italian school-class, and the possibility that immigrants of Islamic origin could themselves organize the field of training and instruction for their own children. School, we should remember, is not just an environment for gaining knowledge and acquiring skills but also the place where values of the context in which one lives and grows are interiorised. Moreover, in this sense Muslim requests for schools and ad hoc education paths risk being interpreted, by those who disagree, as a dangerous attempt to reinforce a particular ethno-religious identity at the expense of a process of progressive integration and positive interaction with the Italian context, which schools naturally favour. Altogether, however, few Italians would be in favour of the institution of schools and specific educational programmes for the children of Muslim families, while the majority of the population is equally divided between those who are wholeheartedly opposed to that solution and those who are half-hearted and unsure about the question. Among the latter, Southern Italians and university graduates predominate, two social groups who are therefore, for different reasons, more tolerant of granting certain academic prerogatives and educational programmes to Muslims. On the other hand, Catholics are not particularly closed on the subject, mostly aligning themselves with the average tendencies of the population as a whole.

Now, we come to another issue, the role of women, which is often used in public debate as proof of incompatibility between Western society and Islam. This is confirmed by our research. The vast majority of Italians find totally unacceptable both repudiation of one's wife in the case of adultery (67.3%) and the duty of girls to wear the veil (58.5%), as against a small proportion of the population (6–7%) who approve of these customs and a minority (30%) who are neutral or open on the questions, half-way between rejection and consent. Of course, what creates problems is not so much the wearing of the veil in itself on the part of Muslim girls as their being obliged to do so, the perception of restraint and forcing of religious traditions considered anachronistic and oppressive towards women, or the revival of a custom considered a sign of subjugation. In this regard, these negative reactions may be tempered by two considerations: above all the fact that women "*with a handkerchief*" are part of our national memory, recalling a rural Italy (which later evolved) in which a veil was a sign of the modesty and reserve required of women at a particular stage of Italian history. Secondly, the meaning that young Muslims attribute to the veil, understood as a symbol—as some scholars have revealed—not so much as a specific identity marker as a means of asserting that it is possible to be Italian without adopting all the customs and traditional usages of Italian society.

Finally, there are two problems connected with the practice of being Muslim, whose solutions may contrast with the organization of Italian social and working life: a particular diet (respecting religious precepts) and the possibility of being absent from places of study and work in order to attend Friday prayers. The need for a special diet is the one most easily accepted by the population, partly because putting it into practice is not seen as a serious problem. Only one-third of Italians still consider it a taboo subject, while 36.1% accept it without reservation and 41.7% do not care or do not consider it a problem. In this connection, we should point out that introducing special diets in schools and hospitals is by now a widespread practice, testifying to how religious pluralism can be seen in everyday life.

The recognition of particular rights during school and working hours is, however, more controversial: such an opening would mean granting "special" prerogatives to a particular religious minority, which could both modify organization of the social system and encourage analogous demands from other faiths in the country. This seems to be one of the topics in the controversy surrounding an agreement between Muslim representatives and the Italian government. Perhaps for this reason fewer than half (46.1%) of Italians are totally opposed to the possibility that Muslims should be absent on their holy day (Friday) from school and work to go to the mosque, compared with a reduced percentage (13.9%) which recognizes that right and a wide band (40%) who express an in-between, more evasive view of the subject. Both those who have no religious affiliation and "by-tradition-and-training" Catholics are less inclined to concede free Fridays to Muslims; while greater tolerance on the matter is expressed by graduates than by those with a low educational level, and by Southern more than by Northern Italians.

The generally greater mistrust among Italians of Muslims in Italy does not prevent them from recognizing a quality of this ever-increasing phenomenon: religious commitment, involvement in terms of faith, which arouses appreciation and reflection. About 50% of Italians admitted that Muslims are more serious about and dedicated to religious practice than Christians are, an opinion not shared by little over one-quarter of the population, while 23.9% said they were unable to answer the question.

This positive opinion of Muslims is quite diffuse in all population groups, even if expressed more by men than women, more by young than old people, more by Northern than Southern Italians, and more by graduates than those with a low educational level: in short, by the groups in the national panorama who are least familiar with religious practice. This in itself is a curious fact. It is, above all, the Italians who least attend religious environments (thereby either showing up their own condition or the lukewarm way the prevailing Catholicism is seen in the country) who declare that Muslims are more religiously involved than Christians are. Christians' religious weakness compared with Muslim faith is also acknowledged by a significant number of the most committed Catholics (those who define themselves as "belonging, active and convinced") yet below the average level for the population as a whole.

However, generally speaking, Italians continue to have a negative opinion of Islam. Data tell us that even though these opinions are not distributed evenly throughout the country: older, less educated people resident in the Northern part of the country are generally less open to Islam. These trends are present not only in Italy, and are well known to scholars of the topic.

However, the regions greatly divide Italian society and culture. Regional boundaries are important for defining not only attitudes towards Catholicism (Garelli 2020), the religion of the majority of the population, but also the acceptance of other religions.

7.4 The Silent Revolution: Children of Religious Immigrants Coming of Age

Although recently overshadowed in the media by the events of the landings in the Mediterranean, the children of immigration—in every migratory process—recall the structural changes typical of every society of arrival of migratory flows. At the same time, they represent a central element of the reflection on the inclusion policies of Italian and European towns and cities. In fact, the second generations are the litmus test of the effectiveness of the integration process: working for (and especially with) them, means—indeed—working for the whole citizenry. From educational policies to youth policies, from inter-generational relations to those among young recruits to be trained and educational figures, in all these cases action is taken to build and/or strengthen processes of civil coexistence within a citizenship that becomes more aware of the transformations induced—and essential—by immigration.

Second, analysing the second generations means deepening reflection on the ability to act of all the actors on the stage of relations among natives, migrants and their sons and daughters (from teachers to associationists and so on).

Becoming an adult in an immigration context (lived personally or transmitted by the experience of parents) involves effort, because one is called, willingly or not, to come to terms with one's past and to take on adult roles when one would like to be only "young". In this elaboration of the relationship with the origins, there is still a lack of educational figures and continuous training offers, just when these young people need answers to important necessities. Those linked to self-respect and the parallel need to be recognised as active subjects of society, a potential resource for the country in which they live. However, above all, the need for social relations and belonging, which can also evolve into a real strategy of identification and integration in a given place.

Educational guidance and accompaniment needs emerge, which—once again—underline that it is time to stop focusing on the personal (increasingly minority) or family component of the migratory experience, emphasizing instead the similarities these young people have with their Italian peers of Italian origin, with whom they share fears and worries about the future, as well as the job-seeking projects that often lead to the need to broaden their gaze towards other nations. The gaze, for everyone, is disenchanted: the Italian (and European) reality in the Mediterranean area is perceived as incapable of offering space and opportunities to young people, even more so if they are foreigners.

Finally, second generations represent an important test for intergenerational dynamics, both within the different ethnic communities and for the whole citizenship, where the different integration paths of the children of immigration challenge the perceptions of adults and already elderly people. Numerous demonstrations of this process, which is already underway, can be found on a personal and social level. For example, young people of foreign origin who arrive at a university degree are, emblematically, the symbol of the success not only of families but also of the work of teachers, parents, animators, and cultural mediators, who, over the years, have built dynamics of positive integration. In the same way, the emergence of associations aggregated through the identity of "second-generation young people" makes it possible to plan and implement extensive awareness campaigns towards a society that is still struggling—after many decades—to acknowledge the children of immigration—their demands and their potential.

7.5 New Generations Challenge the Link Between Ethnicity, Faith and Identity

The growth of a second generation and of the number of those who become EU citizens are two elements that mark a symbolic change and concrete effects within the European scene. A transformation that concerns the direct protagonists (the young) and reverberates in their relationship with society, translates into the relationship with religious identity and the wider debate which strongly links Islam with terrorism and security affairs. The entry into adult life and into the European community of the children of Muslim immigration invites scholars and professionals, service providers, teachers and residents to become aware of the fluidity and the many possible variations in terms of religious identities and to cope with the risks of stigmatisation and marginalization of the youngest generations. In this sense—and also on the side of faith—interviewees give voice to a reality that no longer wants to be in the spotlight because it is foreign or traceable to practices and rituals of other geographical contexts, but to be considered real to the generation of young people who now study at university, seek work, try to become adults and build an autonomy of living and income. In short, even in religious matters, the desire to be "*treated like Europeans and as Europeans*", the desire to "*believe or not, attend weekly or rare functions*", the request to "*stop being children of foreigners, children of immigrants, children of Filipinos or Peruvians. We are boys and girls who live here and we ask to be judged for what we have studied, for what we do, not for what we have inherited, including religion*".

Despite the fact that the religious panorama is increasingly plural, the focus is always on Islam. A polycentric universe, even if stereotypes and simplifications continue to represent it as univocal and devoid of facets. Alternatively, even fractures. Yet there are many. Moreover, in this case, it is a question of recording an increase in the so-called "Islamic archipelago", in the number of prayer centres, cultural associations and mosques in small and large centres. There has also been an increase in the number of reference figures and their degree of interaction with local, national and even European realities. There is concern about the second generation and the perception that European society has of them, of how to guarantee religious socialisation in a context where being a minority confession often means not having adequate space, and still facing closures as happened at the beginning of immigration. The side effect of this situation could be a negative impact on social inclusion: lack of recognition, continuous experiences of discrimination and identity misperception, could reinforce feelings of being marginalised, opening the doors to negative attitudes towards the town/city/country of residence until becoming radicalized.

However, this strong hostility and the need to regain confidence every time after a negative event, shows a difference between Islam and Catholic-Christian religious affiliations. A factor that affects how young people decide to treat their relationship with spirituality, beliefs and rituals and the consequences that derive from them.

> When my Romanian friend told me that women go to Mass with a white veil, nobody reacted. On the contrary, the teacher said that this was the custom in Italy, too. When I said about the veil, a companion of mine immediately said the veil is not acceptable in Italy, that all women should be forced to remove it. Some girls also intervened. I tried to explain what the veil means to us, the idea behind it, the choice, but they would not even let me talk. In the end, I slammed my fists on the table and spoke of the Muslim countries where you do not have to wear the veil, like Tunisia, of the deputies in Morocco, Muslim women who are not enslaved as many think. There is a lot of ignorance and school does not do much. There is little talk of Islam; there is no room for discussion. Even the teachers do not seem to want it. Therefore, Romanian students who do not create problems or those from Catholic countries like the Ukrainian girl who arrived in my sister's class or the Peruvian girls I had. Everything simple, no flashy symbols, no special requests or parties that are not celebrated in Italy. In the end you have to survive: or choose to be stronger and then do as my cousin, she is not afraid, wears the veil, argues, even on the street, when someone is out of line; or, choose my way, be quiet, say nothing and try to be invisible" (Morocco, p. 21).

For Muslims, the focus is on symbols, traditions and even changes in emigration of a religion that every attack near or far puts under observation, reserving to its protagonists—already European citizens or even just permanent residents—the treatment of "*unwanted guests and to be treated with suspicion*". The issue of prejudice often resurfaces and strengthens, demonstrating that migration is still an issue tolerated by many, but introduced as a figure of the new millennium only by a part of the population that, hopefully, does not represent a small minority. It is undeniable—and the data of the most recent surveys on the fears of many Europeans confirms this—that immigration and some of its protagonists (especially those we presume to be Muslims, because they hold the passport of a country where Islam is the majority religion), represent a growing concern.

Today, the need re-emerges for places, activities and projects to develop knowledge, promote debates, update ideas and verify impressions that, if wrong, risk triggering dangerous social conflicts.

Among young Muslims, new priorities emerge. In particular, it is in this phenomenon that differences emerge between the first and second generations, between fathers (and behind the scenes, mothers) and sons (and, not always backstage, daughters). For the former, the infrastructure of prayer and legal recognition has been, since the 1990s, a non-negotiable objective. Today, children seem more interested in building and promoting environments of sharing, developing projects of mutual knowledge, becoming points of reference for institutions, services (from schools to social welfare services) and citizenship thanks to their being a bridge between European society and those who still stand on the margins, between those who delude themselves that they can replicate the environments in which they grew up in a different context and those who fear the neighbour with the veil or the group of men who gather on Fridays in the courtyard of the buildings. Some young people have decided to face this task, partly to witness how being Muslims and Europeans is possible. It is a challenge, however, supported by a few, when compared to the number of children of Muslim immigration who reside on the old continent.

In these scenarios, the role of young people and second generations appears to be a keystone. They are "*sought-after and welcome*" because they represent the guarantee of numbers and integration into society, for their ability to interact with local administrators and supranational organisations and participate in European events.

References

Allievi, S. (2010). *La guerra delle moschee*. Marsilio.

Angelucci, A. (2017). Libertà religiosa e cittadinanza integrativa. Alcune note sul "vivere assieme" in una società plurale. *Stato, Chiese e Pluralismo confessionale, 39*, 1–13.

Berger, Z. (2015). *Altäre der Moderne. Religion in pluralistischen Gesellschaften*. de Gruyter.

CLIP. (2010). *Intercultural policies in European cities*. European Foundation for the Improvement of Living and Working Conditions.

Evans, J., & Baronavski, C. (2018). *How do European countries differ in religious commitment?* Pew Research Center.

Ferrari, S., & Pastorelli, S. (Eds.). (2012). *Religion in public spaces. A European perspectives*. Ashgate.

Garelli, F. (2020). *Gente di poca fede*. Il Mulino.

Kivisto, P. (2014). *Religion and immigration: Migrant faiths in North America and Western Europe*. John Wiley & Sons.

Martínez-Ariño, J. (2017). Conceptualising the role of cities in the governance of religious diversity in Europe. *Current Sociology, 65*(1), 1–18.

Mazzola, R. (2014). Confessioni, organizzazioni filosofiche e associazioni religiose nell'Unione Europea tra speranze diffuse e problemi emergenti. *Stato, Chiese e Pluralismo confessionale, 3*, 1–12.

OECD. (2018). *Working together for integration of migrants and refugees*. OECD Publishing.

Pace, V. (2004). *L'islam in Europa. Modelli d'integrazione*. Carocci.

Penninx, R., Kraal, K., Martiniello, M., & Vertovec, S. (2004). *Citizenship in European cities*. In *Immigrants, local politics and integration policies*. Ashgate.

Ricucci, R. (2021). Managing the Muslim diaspora: Comparing practices and policies in several European cities. In S. Bonino & R. Ricucci (Eds.), *Islam and security in the west*. Palgrave Macmillan.

Soper, J. C., & Fetzer, J. S. (2005). *Muslims and the state in Britain, France, and Germany*. Cambridge University Press.
UNHCR. (2019). *Refugee and migrants arrivals to Europe in 2019*. Retrieved 20 August, 2020, from https://data2.unhcr.org/en/documents/download/70494
Vidal, A. C. (2001). *Faith-based organizations in community development*. HUD's Office of Policy Development and Research (PD&R). Retrieved 1 December, 2020, from https://www.huduser.gov/portal/publications/faithbased.pdf

Chapter 8
Conclusions

European society today is facing immigration as a factor profoundly transforming its communities, a structural element of a society, which is becoming increasingly pluralistic.

The integration of a great number of migrants, particularly second and third generations, is thus becoming an inescapable element in reflection upon schools, work, culture, urban spaces and religion. On the levels of research and public institutions, this calls into question established models, raises (new and old) fears about social solidity and reinforces the heated argument between nationalistic sentiment impulses towards multiculturalism, which forms part not only of parliamentary debate so much as the daily lives of many Europeans (Van Hootegem et al. 2020).

This dialectical relationship—non-linear and not without tension—is constructed above all in those town and city areas where the presence of foreign, or foreign-origin, citizens is most evident. Reduced access to welfare and international crises can undermine laboriously built cohabitation processes. Suffice it to consider management of the increase in asylum-seekers, EU institutions' ambivalence on the subjects of welcoming and cultural recognition, and phenomena of radicalisation and international terrorism.

It is, however, on the ground that the fabric of integration among composite, heterogeneous and plural realities is woven, often through a process of trial and error. If, from a personal point of view, migrants experience transnational biographies with similar elements, it is in social and community relations, often based on complicated economic dynamics, that conditions are created for effective progress in the gradual cohabitation process. To grasp how this proceeds, one needs to watch carefully what happens on a micro-social scale: blocks of flats, neighbourhoods, schools, workplaces, associations and churches are the places where the critical and potential, problematic elements and the hopes of migratory paths can be understood.

The immigration debate, including its rhetorical and excessively over-simplified aspects, may then lead to a positive comparison between distant ideological positions, the themes of migrants as the destructive part of European society (of employment, health resources, social assistance, the property market, the decorum of some

R. Ricucci, *Ethnicity, Identity and Faith in the Current Migratory Crisis*,
https://doi.org/10.1007/978-3-030-84056-3_8

urban areas, identity, Christian-Catholic pre-eminence) and those who consider them, even in times of crisis, as an essential constructive part of the future of Europe.

8.1 Religious Identity in a Globalized World

The religious message enters the daily life of everybody—even of those who declare their distance from any otherworldly faith—by means of shared values, habits and symbols. For example, various athletes have recourse to greater or lesser rituals, or have tattoos reminiscent of religious symbols: many will recall Usain Bolt making the sign of the cross before races, transmitted to the world during the London Olympic Games.

Talking about and discussing religion means moving on a complex, populated and global terrain. Complex, because the religious and spiritual boundaries are less and less clear or uniquely defined. If this is not surprising, it is now equally evident that religion has not lost relevance either in the life of individuals or in the public sphere. The prophecy, on which many would have bet, of the disappearance and reduction of its importance with the advance of modernity (and post-modernity), has not actually come true (Soper and Fetzer 2018). Beyond the wind of secularization, looking at the European scenario and at the numerous debates on religions at various institutional levels, we can comment that '*God is not dead, but he is in good health*'.

It is also necessary not to limit oneself to the scope of Christianity or the West. From Islam to Buddhism, to the new religious and spiritual movements, the theme must now be understood within world and cross-cultural dynamics, also following the significant processes of human mobility of recent decades. These elements are the basis of a recent change of perspective in the sociological view of religion. A generation ago, the dominant paradigm of this voice would have been that of the meaning and consequences of a loss of relevance of the religious message, in both the public space and the private sphere. In a nutshell, the reflection revolved around the consequences of processes typical of advanced modernity that would have reduced—to the point of cancelling—the influence of religion, which would be followed by "*a collective amnesia regarding religious history, beliefs and practices*" (Clarke and Ware 2015, p. 3). Twenty years later, the landscape is completely different: studies and research underline the vitality of the sacred at individual, group and institutional level (Casanova 2009; Norris and Inglehart 2007).

However, caution is required: the debate on secularization has not stopped. Many of its aspects remain to be explored, for example with regard to relations among religion and biology, cognitive sciences, the environment, irreligion, diversity or religious pluralism (Smith 2018). The agenda of interests linked to the study of the sociology of religion is therefore broad and very articulated, aimed at exploring new themes, without however neglecting the classic areas of study, following the dimensions of religiosity: belief, belonging, practice, experience and knowledge (Glock and Stark 1965). These insights have allowed us to investigate how the relationship with the sacred intervenes in orienting everyday life and the actions of citizens in numerous areas: from political to bioethical choices, from values to economic choices, from relations with institutions to numerous and autonomous forms of spirituality; also towards diversity, in all its dimensions (linguistic, religious, political, physical…).

The activity of FBOs in multicultural contexts of various sectors of the economy is crucial. Moreover, it has always been so in the management of catastrophic events such as earthquakes and tsunamis. Who does not remember the intervention of religion-based organizations, from Haiti to Sri Lanka? Then there is intervention in many developing countries and the support of the great international organizations in managing refugee camps in Lebanon, Central Africa and Europe.

The various religious traditions play a fundamental role in welcoming the needy and society's rejects, and thus migrants.

We consider also the emergence of the phenomena of "believing without belonging" and "belonging without believing", referring to those who declare religious affiliation for reasons of ethnic-national identity rather than of faith.

Among various interesting themes, we go into depth on that of religious pluralism and—in the European case—with Muslims. This is not a new focus but becomes central whenever one has to deal with arrivals and migratory waves, forgetting that not only the followers of the prophet Mohammed are on the move.

We are careful not to assume that "religion is only a matter of migrants". Religion is a subject of study and visible among welcomers: operators, volunteers, associations and local administrators are driven by values and ideals with religious roots. Religion is present not only among those who arrive but also among those who are ready, willing and able to face difference—the foreigner linked to a distant God or the same God, but with a different way of praying.

Thus, a multiple vision is necessary—embracing the variegated aspects of natives and migrants, citizens and foreigners, first and successive generations—in order to describe the various ways in which religion interweaves with human mobility. These are not new topics, but they returned centre-stage with the 2015 refugee crisis: religious associations form an active part in welcoming, while right-wing political parties in "besieged" countries appeal to Christian-Catholic roots to erect barriers against the Muslim invader, and marginal youth are attracted by messages using religious language to contrast marginalization and disintegration paths. Then there are those who appear carrying a rosary or a crucifix, and those who seek refuge in mosques or ethnic churches. Alternatively, Pope Francis intervening on behalf of migrants but being contradicted by pro-Orban Hungarian bishops and cardinals. Then there is daily life, the multicultural reality, which, by now, has become normal: its actors are long-term migrants, first and second generations, European citizens, men and women, young and old, religious and atheists, with religious identity visible or hidden for fear of discrimination.

8.2 Not the End of the Story

In building social cohesion, the most significant role is played by people, especially those human resources who perform training tasks. Particular attention must be paid to them. For some of them, working for several months on a project for immigrant children and families is an opportunity to update their "views" of a fast-changing reality such as that of migration and to realize the many facets that can be entailed by being a foreigner. In this direction, intercultural and training activities qualify as useful actions in order to overcome outdated stereotypes. Nevertheless, this is still the experience of only a few people. The task of modernizing the changing European situation must include all realities of the public and private sectors that, depending on the roles and fields of activities and relationships, help define the perception of everyday life. This perception today raises many concerns about the processes of cohabitation of people (young and adults, minors and elderly) of foreign origin, albeit with different bios, characterized, like their peers, by some elements (family traits, educational and citizenship policies, practices of institutional discrimination) in the definition of their life paths. These elements can change, and sometimes determine, how they are perceived and their prospects for social integration. In this context, educational, cultural and religious institutions play a significant role both in accompanying the definition of identity of people of foreign origin and in favouring knowledge paths and analysis of a new reality in the making; thus so.

At this point, it is worth remembering—in tune with key and well-known personalities, from the former US President Barak Obama to the former President of the EU Central Bank, Mario Draghi—that societies have to invest in young people, highlighting their skills to navigate in a multicultural and multireligious world, whatever their family backgrounds are.

The EU increased its focus on diversity and its management at local level, encouraging towns and cities in promoting inter-culturalism at every stage of societal life. The Agenda for Culture in a Globalizing World (2007) defined intercultural dialogue as a tool contributing to the governance of cultural diversity within European societies. This agenda confirmed the importance of the issue at EU level reaffirming principles already stated in the EU Charter of Fundamental Rights (2002), the Racial Equality Directive (2000), and the Employment Equality Directive (2000). Indeed, especially from 2000 up to now, numerous documents (Recommendations, Declarations and Reports) have contributed to enriching the attention paid to inter-culture within the EU framework. These efforts have been supplemented with two other initiatives. On the one hand, the launch of the European Year for Intercultural Dialogue aimed at developing and spreading awareness of the importance of intercultural dialogue at both national and local levels, as already acknowledged in the "Common Basic Principles for Integration" (CBP). According to CBP No. 7, "*frequent interaction between immigrants and Member State citizens is a fundamental mechanism for integration. Shared forums, intercultural dialogue, education about immigrants and immigrant cultures, and stimulating living conditions in urban environments, enhance the interactions between immigrants and*

Member State citizens". This year, the "Intercultural Cities" project was launched, promoting the so-called "intercultural integration model" based on the following principles: emphasizing shared values and a plural identity; adapting governance, institutions and services to a diverse population; desegregation/cultural mixing in institutions and public spaces to build bridges and trust between ethnic communities; dealing with ethnic conflict through mediation and open public debate (CoE 2008). This approach stressed the role of culture and the positive effects of diversity in the development of towns and cities. According to the CoE's perspective, "*rather than ignoring diversity (as with guest-worker approaches), denying diversity (as with assimilationist approaches), or overemphasizing diversity and thereby reinforcing walls among culturally distinct groups (as with multiculturalism), interculturalism is about explicitly recognizing the value of diversity while doing everything possible to increase interaction, mixing and hybridization among cultural communities*" (CoE 2008, p. 27).

Towns and cities in Southern Europe have recognized the growing importance of the immigrant population in their various contexts since the mid-1980s; policies have been developed in order to face the socioeconomic transformations resulting from the immigration process. In the process, one of the most important characteristics and lessons leant emerged: the strong collaboration between institutions and associations, both ethnic and interethnic, cultural and religious, working in the social, education and leisure areas. Therefore, in Italy too, we are not surprised to observe the difficulty Italian politics has in reforming legislation on the citizenship question. In addition, here, it is necessary to underline two apparently contradictory elements, which are united, however, by their degree of importance in the long term. First, the path to effective recognition of the right to a passport is a fundamental factor in second-generation integration, and consequently an increasingly important element in encouraging this component in Italy's (still-young) experience as an immigration country. Second, this is only one side of the coin of relations between ethnic communities and local society, which are woven in a complex, and sometimes contradictory, fabric of relations among natives and foreigners, old and new residents, and successive generations. From time to time, numerous challenges have emerged asking local administrations to cope with the:

- continuance of building social cohesion in a difficult economic period: the rate of unemployment is rising with the number of Italians looking for work;
- development of intercultural abilities among social servants, teachers, citizens, Italians and foreigners;
- promotion of knowledge of the various facets of the immigrant population: from the first generation to the second; from Muslims to Orthodox; from cleaners to ethnic entrepreneurs; from people with the right to vote to asylum seekers;
- most important lesson emerging of the importance of a network of associations and institutions working together to define and implement integration in an intercultural way. As an educator said, "*We don't have a specific intercultural policy: we are intercultural in our way of working, in our activities, in our mentality. We organize the celebration of the end of Ramadan as well as Easter Mass. We*

discuss the difficulties of growing up as immigrant adolescents as well as discrimination in access to work. The main characteristic of our job is to <work with>, not 'work for'".

Judging from results, the approach taken has numerous strong points:

- A bi-dimensional approach: initiatives are numerous and various, focusing not only on the reception time span but also on the subsequent integration itself;
- Attention to improving the involvement of immigrant/ethnic associations in general projects not only addressed to single minorities (e.g. Chinese) or specific groups (e.g. Muslims);
- A view at sub-local level: focusing attention on the factual implementation of strategies and not only on their concept/promotion, allowing the municipality to understand better the difficulties of immigrant associations and administrators when defining strategies, interventions and policies within the societal area;
- The definition of interdepartmental meetings and projects: a comparison with workers coming from different fields enable the municipality to understand better the potentialities and shortcomings of the strategies implemented in various societal fields;
- The involvement of NGOs working with immigrants, allows the municipality to investigate better the needs and problems of foreign people and to evaluate the proposed strategies;
- It is important to continue to support programmes concentrating on the following core areas: language acquisition, education, integration into the labour market, social counselling and social integration;
- Nonetheless, the integration programme is not a one-time project but a continuous, dynamic process open to address other issues as they become important. In this way, projects intended to promote civic education, participation and solidarity among the second generation represent a willingness to think beyond current problems and aim towards the future.

8.3 The Strength of Weak Links

Inclusiveness is its fundamental characteristic. The aim is to help, educate, entertain, perhaps direct, express, all those who agree to collaborate.

The process that, from the analysis of a territory and its protagonists, favours the knowledge of it, which represents the keystone in order to be able to play the role both of those who welcome and those who are welcomed. For the former, it is a matter of *understanding* (and *learning*) what the characteristics of the newcomers are, as well as which tools can be useful in order to be able to *define* (and therefore *act in*) paths of accompaniment, sustainment and support for a positive social integration. For the latter, on the other hand, it means drawing the boundaries within which to learn how to manage and handle the tools useful to face their destiny as young people that the experience of migration has sometimes deeply felt.

All this within a socio-economic environment in transformation, just as in the towns and cities of Italy and southern Europe, recent immigration countries and, since 2014, put to the test by the emergency of asylum seekers and refugees along the Mediterranean and Balkan routes. The average age of those arriving is low—many are minors or young people under 30. The educational field is where the difficult game of reception and the reduction of differences in integration paths, social cohesion and intergenerational support is played. In other words, one learns to become an adult, putting into practice knowledge learnt at school and values introduced into distant family environments, which have roots, traditions, values and customs in different contexts.

The task facing local administrations and NGOs, and ethnic and religious organisations, is therefore not only to welcome, but also to offer opportunities for those who arrive and those who must learn the new grammar of multicultural societies. In order to meet this challenge it is necessary "*to adopt as a theoretical reference framework the solidarity among the subjects of training and education, and not [...] competition and rivalry. As a first step, a strategy of solidarity implies the precise recognition of existing diversities in order to be able to develop tailor-made methods of intervention, adapted to the characteristics of the needs of each individual subject included in the network*" (Ricucci 2014, p. 78).

What is at stake is social cohesion, in a country that knows how to develop inclusion processes. For everyone. Natives and migrants, citizens and foreigners. Men, women, adults and children: all contribute to a nation's human capital endowment. In fact, no country needs only excellence to grow economically and civilly, but also, and above all "*a fabric of widespread skills*".

In this framework, associationism plays a great role: being one of the pivots of the educational community that educates people to live together and promotes respect and tolerance. This construction can only take place at local level. Furthermore, here is another element that complicates the picture: the proximity of social transformations. The challenge of managing the change triggered by migration takes place, therefore, in neighbourhoods, towns and cities. If at neighbourhood level (and its educational actors), the daily reference area, it is the task of teaching (in a formal and informal way), to decipher the changing reality (from the social point of view, of education and training paths, of work); at town and city level, it is the task of developing policies to support young people to recover the protagonism they have long lost.

The characteristics of each "*integration site*", which wants to go beyond the first reception and the only response to basic needs, are ideally synthesized in three words: *Reception, Openness and Acceptance* of the other. It may be useful to propose an example for each aspect.

Reception is that which allows migrants to realise their potential and be recognised as people with a history and not a '*tabula rasa*' to be coloured in the image of the welcoming community.

Openness recalls the willingness of the operators to get involved and go back to school. It may seem a recommendation that transcends those who work in associationism, in FBOs, and organisations that by definition should be flexible, open to change, to diversity. Yet, even the world of volunteering contains rigidities and rivalries that undermine the effectiveness of interventions. Sometimes, it is actually more difficult to welcome from a religious and ecumenical point of view: "*everyone is welcome, as long as they pray and believe as we do*", is one of the refrains that emerged in interviews with Catholic or Protestant faithful in front of co-faithful from other countries. Moreover, to think that history reminds us of similar comments made by the Irish against the Italians (both Catholic groups) in Melbourne, in the 1950s.

Finally, *Acceptance* of the other can only bring to mind the continuous work of weaving what is built daily through activities, meetings (and some clashes), opportunities to promote understanding and acceptance of diversity, not only of colour or religion, but also of language, social condition, generational belonging. In this sense, common initiatives of civic participation are attempts to promote the encounter with diversity.

Ideally, three steps that refer to a positive-intercultural attitude define the style of those who have been met to write this volume. At all levels. Between salaried and voluntary workers; between workers and recipients; among different generations; among distant origins. It is often thought that the intercultural environment represents a sort of "*free zone*", where stereotypes and prejudices are broken down. This is not the case, of course. It is actually the outcome of work, first and foremost, which everyone carries out, on themselves, also through the training and information activities that the structure offers.

It should be remembered that, at the end of this path, the real test of today's insertion processes will take place in the coming years, when those who have arrived, especially those who are minors or just come of age, will be confronted with the world of work, the construction of lasting affective relationships, the instances of political participation in the areas where they live. The challenge then becomes that of not considering the projects and successes achieved so far as a point of arrival, but as a stage within a (more or less long) process of inclusion. This process has its most qualifying points in the active involvement of actors within a territory, and in the enhancement, above all, of young people and their energies, in a process of building active citizenship and in the interweaving of different skills. In other words, in the characteristics of tomorrow's citizens and in how citizenship in general will be able to react and measure themselves against a reality that often has different traits from those in which they grew up. It will also depend on the sustainability of policies aimed at promoting the (not only subordinate) integration of immigrants.

8.4 Final Thoughts

Therefore, although I agree that there are powerful global trends at work in the book's topic, I should not lose sight of crucial differences in how those trends are experienced and reshaped from one place to another. For this reason, I have hesitated to reduce my findings to an example of a new season of religious engagement.

Additionally, combining everything under one label risks losing sight of the variations that still characterize Southern European towns and cities and the others across Europe. However, the forces that encouraged the use of the Italian case as peculiar and representative of the current mood experienced in the recent immigration countries and those re-effected by the long shadow of the refugee crisis, led to a remarkable transnational and trans-religious cross-fertilization in the repertoire of experiences made and activities taken. Exploring and analyzing findings and their limits has brought me into dialogue with several large theoretical trends.

First, transnationalism is an important factor in the limited sense; people, even in the digitalized era and in the most powerful period for the spreading of internet, people are strongly locally rooted. They are, of course, exposed to the wind of globalization; however, their behaviour is more conditioned by their close (real) environment than the virtual and far away ideas. The wind of secularization blows strong and atheism is increasing among young people belonging to all religious denominations; however, people still continue—sometimes unconsciously—to experience values, acts, and places dealing with religions. Working at or volunteering for Caritas, helping migrants to find a place for worshipping, enjoying Pope Francis' openness towards migrants, being an Islamic Relief supporter, promoting intercultural and interreligious activities, and sustaining FBOs are some examples.

Second, national history and memories do not work explicitly. Indeed, when native-residents support charities or open their homes to welcome migrants, recalls some traits of the experiences of Italians, Spanish, Portuguese and other Europeans, back at the beginning of the twentieth century in various countries. Interaction with history and the idea of a kind of policy, learning from the past is theoretically solid, but disconfirmed practically. The findings in the book confirms that local socio-political contexts are still crucial units of power, not just shaping the religious and organizational, but providing the grounds for the development of projects that are, in some way, interacting with religions.

Finally, migration continues to solve the "mirror function': analyzing the reality through the lens of migrants and their descendants, still means highlighting our discrepancies, un-solved issues in our own societies. Moreover, the role of religion in lives, economy, political debate, socio-cultural environments is under discussion: migrants in some way help host societies to unveil their lies (religions and faiths are only personal matters) and to deal with the current and future religious pluralism.

References

Casanova, J. (2009). Immigration and the new religious pluralism: A European Union – United States comparison. In G. Brahm Levey & T. Modood (Eds.), *Secularism, religion and multicultural citizenship* (pp. 139–163). Cambridge University Press.

Clarke, M., & Ware, V. A. (2015). Understanding faith-based organizations: How FBOs are contrasted with NGOs in international development literature. *Progress in Development Studies, 15*(1), 37–48.

CoE. (2008). *White paper on intercultural dialogue*. Council of Europe.

Glock, C. Y., & Stark, R. (1965). *Religion and society in tension*. Rand McNally.

Norris, P., & Inglehart, R. (2007). *Sacred and secular. Religion and politics worldwide*. Cambridge University Press.

Ricucci, R. (2014). *Second generations on the move in Italy*. Lexington Books.

Smith, C. (2018). *Religion*. Princeton University Press.

Soper, J. C., & Fetzer, J. S. (2018). *Religion and nationalism in global perspective*. Cambridge University Press.

Van Hootegem, A., Meuleman, B., & Abts, D. (2020). Attitudes toward asylum policy in a divided Europe: Diverging contexts, diverging attitudes? *Frontiers in Sociology, 5*, 35. https://doi.org/10.3389/fsoc.2020.00035